The Pristine Coast

Other Books by Scott Renyard

Illustrated Screenplays

Who Killed Miracle? (2022)

The Unofficial Trial of Alexandra Morton (forthcoming 2023)

Trial of an Iconic Species (forthcoming 2023)

Children's Books

The Flag That Flew Up (2021)

NO FISH, NO LIFE.

The Pristine Coast

an illustrated screenplay

SCOTT RENYARD

juggernaut CLASSICS

Copyright © 2023 Juggernaut Classics Inc.

The Pristine Coast documentary copyright © 2014 by Salmon Productions Ltd.

No part of this publication may be reproduced or transmitted in any form or by any means, electronic or mechanical, including photocopying, recording or by any information storage and retrieval system, now known or to be invented, without permission in writing from the publisher.

Every effort has been made to confirm the factual accuracy of the information in this book and to identify copyright holders of photographs and obtain their permission to use the photographs. If you are aware of any oversights or errors in this respect, please contact the publisher so that they can be corrected in future reprints or editions of this book.

Published by Juggernaut Classics Inc.
Contact: scott@juggernautpictures.ca

ISBN: 978-1-998836-20-8 (softcover)
ISBN: 978-1-998836-21-5 (eBook)

Cover artwork courtesy of Juggernaut Pictures Inc.
Herring photo courtesy of Alexandra Morton
Edited by Lesley Cameron
Cover design by Michael George and Jan Westendorp
Book design by Jan Westendorp/katodesignandphoto.com

Lyrics for "The Pristine Coast" reproduced courtesy of
Richard Harvey Hagensen, Rene Joanne Banks, and Scott Renyard
Lyrics for "The Get Out Migration" reproduced courtesy of
Richard Harvey Hagensen and Rene Joanne Banks
Lyrics for "The Best Disrespect That Money Can Buy" reproduced
courtesy of Richard Harvey Hagensen and Rene Joanne Banks
Lyrics for "I Am The Future" reproduced courtesy of
Holly Arntzen and Kevin Wright
Lyrics for "Up Your Watershed!" reproduced courtesy of
Holly Arntzen and Kevin Wright

Juggernaut Classics Inc.

I dedicate this book to my dad,
Tom Renyard, for all the wonderful
fishing trips that made me fall
in love with the environment
in general and especially fish.

Contents

Introduction	ix
Cast of Characters	xvii
The Pristine Coast—an illustrated screenplay	1
Acknowledgements	323
Bibliography	327

Introduction

I often get asked where I get my ideas for films. I guess the answer is anywhere and everywhere. Most of the time a chance event of some kind sets things in motion. I might read an article in a newspaper that catches my attention, or I meet someone who has a story to tell, or I see something out in nature. For this film, it was a combination of all three.

In 2002, I was at the Vancouver Aquarium interviewing Dr. John Ford, a world authority on killer whale vocalizations, for the film *Who Killed Miracle?* I was hoping he would be able to determine what pod Miracle, the eponymous star of that film, came from by analyzing a few vocalizations I had of her while she was being rehabilitated at the Oak Bay Beach Hotel. As it turned out, Miracle was still speaking in whale baby talk. So, the analysis was inconclusive. But at the end of the interview, in walked Alexandra Morton. Dr. Ford introduced us and mentioned that Alexandra was a volunteer at Sealand of the Pacific when Miracle was performing at the park. On the spur of the moment, I asked her if she would like to be interviewed and talk about Miracle. She subsequently gave a terrific and revealing interview for the film.

In 2003, I was fishing one of my favourite rivers for salmon near my home. The Chilliwack River, also known as the Vedder River, is home to all five species of Pacific salmon—coho, chinook, chum, pink, and sockeye—and three species of trout—sea-run steelhead, bull, and cutthroat. It's a sport fisherman's dream, even though it receives a tremendous amount of fishing pressure. The Chilliwack River hatchery supports many of the species in the river system and raises and releases coho, chinook, and steelhead fry into the system to support the species targeted by sport fishermen. Another hatchery, the Cultus Lake hatchery, also raises the endangered Cultus Lake sockeye and releases them into a tributary of the Chilliwack River called the Sweltzer River.

In spite of the addition of hatchery fish, declines in some salmon runs were very noticeable as early as the 1990s. Runs of chum and pink salmon, once so abundant, were dropping quickly. The Cultus Lake sockeye population is considered endangered. It was glaringly obvious to many fishermen that something was not right. Although many formal studies were conducted, in reality none were needed to confirm what everyone was witnessing: There were simply a lot fewer fish returning from the ocean to spawn.

The Chilliwack River system is part of the Fraser River watershed, and sport fishermen I met on the river were quick to blame commercial and First Nations net fishery openings downstream on the Fraser River. But those openings had been going on for decades, so this sudden, obvious change in returns to the Chilliwack system was unarguably odd. At first, many people speculated that the Department of Fisheries and Oceans had succumbed to pressure from the commercial fishermen and was allowing the commercial river fishery to remain open too long and overexploiting the runs. Then sport fishermen started blaming each other for mishandling the fish they were releasing. One day, I was shocked to see dozens of fresh-from-the-ocean silver coho and chinook salmon floating belly-up down the river. I initially bought into the rhetoric being exchanged on the banks of that river that there was a new breed of sport fishermen who had no clue how to release the wild fish they were not allowed to keep. But then I noticed many of the dead fish were hatchery fish. Why would an angler mishandle a hatchery fish (identifiable by its clipped adipose fin) and toss it back when they could keep it?

In 2004, I heard a news report featuring Alexandra Morton. She had reported a few years earlier that open net fish farms were causing sea lice populations to explode and were the reason behind large sea lice loads on chum and pink salmon juveniles near her home of Echo Bay, BC. The theory was that the sheer magnitude of the mechanical damage to the young salmon's skins was enough to kill them.

Alexandra's thoughtful observations about Miracle's behaviour made me believe that her concerns about fish farms had merit. So I gave her a call. I asked her, "Do you think this could be affecting fish populations as far south as the Fraser River?" She said that the impact appeared to be more widespread than originally thought. It was a bit mind-boggling to think that fish farms a few hundred kilometres north of the Fraser River could be having an impact so far away. But when I thought about it a little more, I realized that many of the Fraser River salmon populations migrate north through the Inside Passage and past Alexandra's home in the Broughton Archipelago and the fish farms. So, it was not outside of the realm of possibilities that Fraser River fish would get infected with sea lice from the farms. And if it turned out to be true, it would be a huge story. But why were the adult fish dying so far away in the rivers?

The sum of these events, combined with my passion for fish, convinced me I had found my next film project. But after making some initial inquiries, I discovered that two other film makers, Twyla Roscovich and Damien Gillis, were both already working on films about the issue. I decided to give them space and shelved my idea. In the meantime, I continued to go to the Vedder River to fish, and I saw the salmon runs continue to shrink. Fishermen were also reporting that the flesh of some of the fish they were catching would dissolve during the cleaning process or that the fish were full of parasites. I was feeling very guilty that I hadn't pressed on with my idea to make a film about this issue. It was becoming very evident that there was more going on than just physical injury from sea lice.

In 2010, I heard that Alexandra was planning a protest called the Get Out Migration in the form of a walk down Vancouver Island in May. It sounded like there was tremendous energy surrounding the idea, and I knew that it could provide a great spine for a film. I called Alexandra and told her I was eager to jump in and film as much of the protest as possible. She agreed to do an interview and introduced me to the researchers at the Salmon Coast Field Station, which she founded. I found out that they were doing a lot of groundbreaking research on sea lice infestations impacting juvenile salmon passing the farms.

I hired my friend Mark Noda to be my cameraman, and we headed up to Echo Bay to conduct interviews and collect footage the week the protest was scheduled to begin. This was the first step in a journey that would last for four and a half years.

The Get Out Migration protest began at Sointula, BC, on April 23, 2010, and ended at Victoria, the provincial capital, on May 8, 2010. The protesters walked down Vancouver Island, covering over 400 kilometres in the process. The walk started with a small group but grew in size as the protest gained momentum over 16 days. I filmed eight of the protest days—in Echo Bay, Sointula, Quadra Island, Campbell River, Comox, Nanaimo, Duncan, and Victoria—making trips back and forth to Vancouver Island as I juggled the demands of my day job with the new project.

At first, I thought this film was going to be about the impact of open net pen fish farms on Pacific salmon populations. I thought the fact that the farms could be impacting salmon coast-wide, and not just in the Broughton Archipelago where Alexandra discovered the problem, was already a massive story. But while I was filming near Campbell River, my story took a very surprising and dramatic turn. One day, while we were on a break, Alexandra showed me a photograph of a herring that was bleeding from its eyes, from the creases of its fins, and around its face.

I graduated from the University of British Columbia in 1986 with a Master of Science degree, and suddenly all those years of study took on a new and urgent value. The herring photo immediately made me ask, If salmon and herring were being affected, what other species were involved?

I began digging through Alexandra's archives and tracing references to more and more documents. I went to the office of the Fisherman's Union and read through decades of articles published in the *Fisherman* newspaper about shocking and unexplainable changes in wild fish populations. A freedom of information request by Alexandra produced a treasure trove of information that revealed the many concerns provincial and federal bureaucrats had about the introduction of diseases from Europe to BC's Pacific coast. There were dozens of memos and letters, written over more than a decade, outlining what the introduction of exotic pathogens could do to very valuable North Pacific salmon populations. The writers of the letters and memos stated that the introduction of exotic diseases as a result of the importation of live Atlantic salmon eggs was not an "if" but a "when." In particular, they were concerned about the introduction of viral hemorrhagic septicemia (VHS), whirling disease, and infectious pancreatic necrosis (IPN). One senior bureaucrat wrote that it was like playing Russian roulette with an extremely valuable resource, namely Pacific salmon populations. He was not wrong.

Soon a pattern began to emerge: Species after species of wild fish had declined in numbers since the introduction of open net pens on the Pacific coast. Steelhead trout, coho salmon, chinook salmon, pink salmon, sockeye salmon, chum salmon, bull trout, sea-run cutthroat trout, white sturgeon, Pacific smelt, Pacific herring, ling cod, and many groundfish species all began declining in the late 1970s.

It turns out that VHS is a particularly deadly virus that infects just about every species of fish. So my film suddenly became a project that

DR. CRAIG ORR:
: Executive Director, Watershed Watch Salmon Society

BRAD HOPE:
: Former Owner, Tidal Rush Farms

GREGORY MCDADE, QC*:
: Managing Partner, Ratcliff and Company

GEOFF MEGGS:
: Former Editor, the *Fisherman*

ALLAN GUDBRANSON:
: Commercial Fisherman

ANISSA REED:
: Artist and Wild Salmon Activist, Sointula, BC

DR. PETER PEARSE:
: Commissioner, Turning the Tide: A New Policy for Canada's Pacific Fisheries, and Professor Emeritus, the University of British Columbia

OTTO LANGER:
: Retired DFO Fisheries Biologist

DR. JOHN VOLPE:
: Principal Researcher, SEG Group University of Victoria

DAVE GILLESPIE:
: Former Chairman, BC Finfish Aquaculture Inquiry

JOHN CROSBIE, QC:
: Former Minister, Fisheries and Oceans Canada

JOHN CUMMINS:
: Former Canadian Member of Parliament

JOHN FRASER, QC*:
: Former Minister, Fisheries and Oceans and Speaker of the House of Commons

DR. DANIEL PAULY:
: Professor, Fisheries Centre, the University of British Columbia

BOB MCCLELLAND:
: Former Cabinet Minister, Province of British Columbia

GORDON WILSON:
: Former MLA, Powell River-Sunshine Coast Riding

MAC RICHARDSON:
: Resident of Sechelt, BC

BARTLETT NAYLOR:
: Consultant, Corporate Governance

DR. MARTIN KRKOSEK:
: Assistant Professor, Department of Ecology, University of Toronto

IAN GILL:
: Journalist

JUNE HOPE:
    Former Owner, Tidal Rush Farms

BILLY PROCTOR:
    Commercial Fisherman, Echo Bay, BC

DON STANIFORD:
    Anti-Salmon Farming Activist

DICK HARVEY:
    Film Maker, Former DFO Employee

BOB MCILWAINE:
    Former Manager, Fisheries Development Division, Department of Fisheries and Oceans

DARREN BLANEY:
    Chief, Homalco First Nation

HOLLY ARNTZEN:
    Singer

KEVIN WRIGHT:
    Singer

WILLIAM ROUTLEY:
    MLA, Cowichan Valley, BC

TWYLA ROSCOVICH:
    Environmental Film Maker, Campbell River, BC

DR. ALEJANDRO BUSCHMANN RUBIO:
: Professor, Marine Biology, University of Los Lagos

INKA MILEWSKI:
: Marine Biologist, Conservation Council of New Brunswick

FRANÇOIS PERREAULT:
: Sport Fisherman

DR. TRISHA ATWOOD:
: Food Web Ecologist, University of British Columbia

DR. PIETER TANS:
: Senior Scientist, NOAA Earth System Research Laboratory

The Pristine Coast
—an illustrated screenplay

By
Scott Renyard

Alexandra Morton's Get Out Migration protest started in Sointula, BC, on April 22, 2010, and ended in Victoria, BC, on May 8, 2010. Protesters are seen here on a street in Victoria. (Photo credit: John Preston)

The Juggernaut logo opens the picture.

OVER BLACK

SUPERSCRIPT: May 8th, 2010

Drum beats in the distance.

> WOMAN (O.S.)
> (faint, barely audible)
> Wake up, Canada! Wake up, Canada!

The sounds of footsteps running, hard breathing. We can hear shouting in the distance, cop car whoops, drum beats, chanting, horns honking.

EXT. VICTORIA STREET — DAY

CAMERA POV

A low angle of the street, erratic and shaky. The cameraman is running, his sound guy following with his gear.

NEW SHOT

A large group of protesters make their way down the street away from camera. A cop walks with them. Another one drives by on a motorcycle.

                    WOMAN (O.S.)
                 (barely audible)
            No more fish farms!

Her voice grows louder, and so do the drums.

                    WOMAN (O.S.)
            No more fish farms! No more
            fish farms!

EXT. NEXT BLOCK - MOMENTS LATER

The protesters are coming right at camera.

                    WOMAN (O.S.)
            Wake up, Canada!

                    OTHERS (O.S.)
            Wake up, Canada!

A woman dressed in traditional clothing comes into view. She's leading the chant.

                    WOMAN
            NO MORE FISH FARMS!

                    OTHERS
            NO MORE FISH FARMS!

She walks by camera.

                    WOMAN
            NO MORE FISH FARMS!

>   OTHERS (O.S.)
>   NO MORE FISH FARMS!

IN THE CROWD

The woman continues her chant.

OUT OF THE CROWD

A different woman emerges from the group of protesters. She's wearing a bandana and carrying a small flag. It's Alexandra Morton.

She spins around and takes in the size of the crowd, smiling.

>   SINGERS (O.S.)
>   Down, down, do my roots go
>   down.
>
>   Down, down, do my roots go
>   down.

ON THE BC LEGISLATURE BUILDING

The protesters are gathered in front of the steps. It's a large crowd.

>   SINGERS (O.S.)
>   Down, down, do my roots go
>   down.
>
>   Down, down, do my roots go
>   down.

THE CROWD

The protesters shake their signs to the beat of the song and drums.

CLOSER

A sign reads "We want wild salmon."

DRUMS

Take over and the singers fade. The crowd gets louder, bouncing their signs to the beat of the drums.

TIGHT ON AN INDIGENOUS WOMAN

TITLE: Joan Phillip, Council Member, Penticton Indian Band.

It's Joan Phillip, a council member of the Penticton Indian Band and wife of Grand Chief Stewart Phillip. She's emotional.

     JOAN PHILLIP
  My youngest grandson turned one
  today, but we're here, and I
  know that all of you have busy
  lives as well. So I just honour
  you for taking the time to be
  here today, ah, to protect the
  salmon. It's a matter of our
  collective survival.

FROM THE TOP OF THE STAIRS

A huge crowd reacts enthusiastically to Joan's alarming words.

ON FORMER GRAND CHIEF STEWART PHILLIP

TITLE: Stewart Phillip, President, Union of BC Indian Chiefs.

He's at the microphone.

> STEWART PHILLIP
> And it's that intuition that has brought us from many different places . . .

ON THE CROWD

A man holds up a sign that reads "SOS Save our Salmon."

> STEWART PHILLIP (O.S.)
> . . . from all points of the compass, from out of the comfort of our homes and our families,

BACK TO STEWART

> STEWART PHILLIP
> . . . that has brought us here today to serve notice on the provincial government, on the government of Canada, on the Norwegian corporations and
> (MORE)

STEWART PHILLIP (CONT'D)
those other corporations that
are wreaking havoc in the
pristine waters along the coast
and imposing their absolutely
toxic and repugnant fish
farming industry on us.

The crowd cheers, claps and pounds drums.

STEWART PHILLIP
There was a prophecy that the
time would come when the four
races of mankind would come
from the four directions.

ON ALEXANDRA

She listens to his words. And then looks to the crowd.

ON THE CROWD

They listen intently to Stewart's message.

STEWART PHILLIP (O.S.)
And they would come together
because Mother Earth was
in peril. And I believe, I
absolutely believe . . .

BACK TO STEWART

Grand Chief Stewart Phillip of the Syilx Nation and his wife, Joan Phillip, speak at the conclusion of the Get Out Migration protest. (Photo credit: John Preston)

                    STEWART PHILLIP
. . . that that time is upon us now.

The crowd claps and cheers.

EXT. LEGISLATURE LAWN — MOMENTS LATER

Everyone is jammed tight on and against the stairs of the legislature.

OVER THE CROWD

TITLE: Rafe Mair, Political Commentator, Former BC Minister of Environment.

                    RAFE MAIR
          I was thinking today how lucky
          British Columbia is to have
          had such a wonderful effort,
          an unbelievable effort, by
          someone who came here from
          the United States and people
          who have been here forever. To
          think that Alexandra Morton and
          our gallant First Nations have
          brought us to this point is a
          credit to them, and I think we
          should acknowledge it.

Alexandra and the crowd clap and cheer.

                    RAFE MAIR
          I want to say something to
          the people who inhabit this
          building a lot — as I once did.
          You bastards don't own those
          fish! You don't own those
          rivers! You don't own that
          environment! That's ours!

REVERSING

The thousands of people in the crowd cheer
and wave their signs in support. Others beat
drums.

                                    FADE OUT:

FADE IN:

TITLE: Juggernaut Pictures presents . . .

EXT. PACIFIC COAST — DAY

A magnificent coastal mountain rises up out
of the sea. But there is a flaw in the image.
There is a partially overgrown clear-cut
on it.

TITLE: The Pristine Coast

SONG: "The Pristine Coast" begins.

EXT. COASTAL SHORELINE — DAY

Another aerial view. This one looks a bit more pristine.

> RICH HAGENSEN (V.O.)
> Oh, they call it the Pristine Coast. Where the endless tall timber is green.

EXT. SONORA ISLAND — DAY

An aerial shot looks down at two fish farms on a brilliantly sunny day.

TITLE: Written and Directed by Scott Renyard

> RICH HAGENSEN (V.O.)
> Mountains and islands shine blue in the sun. And the rivers flow clear and clean!

EXT. OCEAN — DAY

The sun reflects in the water next to a bloated dead seal.

EXT. BOUNDARY BAY — DAY

A DFO hovercraft with divers on it is next to a dead killer whale. The divers jump into the water with a rope.

13.

> RICH HAGENSEN (V.O.)
> The killer whales play in the
> ocean, in the sparkling water
> that teems . . .

AT THE SHORE

The killer whale has been dragged onto shore.

EXT. STREAM — DAY

Dead salmon are strewn around in the water like sticks from a debris torrent.

TITLE: Director of Photography Mark Noda

> RICH HAGENSEN (V.O.)
> . . . with salmon and shellfish,
> seaweed and seals. But all this
> is not what it seems.

INT. STREAM — DAY

A dead salmon, belly-up, floats slowly downstream.

EXT. CLEAR-CUT (AERIAL) — DAY

A large patch of forest has been freshly cut.

> RICH HAGENSEN (V.O.)
> On the backs of those pristine
> mountains . . .

EXT. FORESHORE — DAY

A helicopter drops some trees into the ocean.

                    RICH HAGENSEN (V.O.)
          . . . clear-cuts are hidden
          from view.

TIGHTER

The helicopter drops a couple of logs into the ocean. And again. Freshly cut logs make up a loose log boom.

INT. OCEAN — MOMENTS LATER

Logging debris and cables choke the bottom of the ocean.

                    RICH HAGENSEN (V.O.)
          Logging roads carve up the
          valleys!

EXT. OCEAN — DAY

Dead salmon lie in the water.

EXT. HILLSIDE — DAY

Power lines slice across a snow-capped peak overhead.

                    RICH HAGENSEN (V.O.)
          Miles of transmission lines
          too. The water round fish farms
          and pulp mills . . .

EXT. FISH FARM — DAY

A large nondescript fish farm floats along a shoreline.

EXT. OCEAN — DAY

A mort barge loaded with totes of dead fish is placed a few hundred metres from a farm.

> RICH HAGENSEN (V.O.)
> . . . is thick with toxic pollution!

TITLE: Narrated by Kirby Morrow

INT. OCEAN — DAY

The bottom of the ocean under a fish farm is devoid of life, except for a pair of sea anemones.

> RICH HAGENSEN (V.O.)
> Infecting the migrating salmon . . .

INT. FULTON RIVER — DAY

A large pre-spawn die-off is occurring at a counting fence. Hundreds of salmon are belly-up.

EXT. HARRISON RIVER — DAY

A dead silver sockeye with its eye picked out floats down the river.

                    RICH HAGENSEN (V.O.)
               . . . and it's all just a
               pristine illusion!

EXT. EMPTY MORT BARGE — DAY

A sign posted on the barge reads "Restricted Area: Authorized Vessels and Personnel Only." It stakes turf in a public ocean.

TITLE: Edited by Maja Zdanowski

                    RICH HAGENSEN (V.O.)
               Ruin of the rivers and fish
               farms,

EXT. HARRISON RIVER — DAY

Many silver sockeye float down the river. Dead.

                    RICH HAGENSEN (V.O.)
               . . . kill everything living in
               sight!

TIGHTER

A bloated silver sockeye floats in the shallows, partially split open from the gas building up in its abdomen.

                    RICH HAGENSEN (V.O.)
               They're gonna keep wreckin' it
               all,

UNDERWATER

A small fish lies dead on the bottom of the Harrison River.

>   RICH HAGENSEN (V.O.)
>   . . . unless we make them do
>   what's right!

EXT. VEDDER RIVER — DAY

Four chum salmon, dead.

EXT. HARRISON RIVER — LATER

Another silver sockeye, dead and bloated with its roe spilling out on the ground.

EXT. FISH FARM — DAY

Atlantic salmon swim around inside one of the cages.

>   RICH HAGENSEN (V.O.)
>   Do we want fish caged in drugs,
>   using dyes that make their
>   flesh pretty?

INSIDE THE NET PEN

Atlantic salmon swim around with their dorsal fins jutting above the water line.

EXT. FISH FARM (AERIAL) — DAY

Two rows of cages are busy with activity.

TITLE: Original music by Stu Goldberg

> RICH HAGENSEN (V.O.)
> Stuff it in 'til they can't eat a bite.

ANOTHER FISH FARM

It sits tucked away in a cove at dusk.

> RICH HAGENSEN (V.O.)
> Dress 'em up for the folks in the city.

TWO MORE FARMS — With two rows of net pens.

> RICH HAGENSEN (V.O.)
> Save the fish and bears and trees, from the frontier mentality kind . . .

INT. OCEAN — DAY

A plastic chair, caked in crud, lies on the bottom of the ocean.

INT. STREAM — DAY

A badly mottled female sockeye salmon swims against the current.

                    RICH HAGENSEN (V.O.)
                . . . with that pristine
                coast façade, and the mass
                destruction behind.

INT. OCEAN — EVENING

A wheelbarrow sits under a fish farm. The
sight is eerie. Marine snow floats around it
like snowflakes falling from the sky.

                    RICH HAGENSEN (V.O.)
                Pristine coast façade, and the
                mass destruction behind.

INT. OCEAN — DAY

A sea lion lies dead in the net of a fish
farm.

INT. OCEAN — DAY

The bottom of the ocean is almost devoid of
any sign of life, except for a single strand
of eelgrass.

                    RICH HAGENSEN (V.O.)
                The pristine coast façade and
                the mass destruction behind.

EXT. FISH FARM — EVENING

An empty and closed fish farm sits in the
ocean. Eerie.

UNDERWATER

A bright silver sockeye, belly-up, floats away and fades into the green water.

                                        FADE OUT:

FADE IN:

OVER BLACK

EXT. OCEAN (1970s) — DAY

A commercial fishing fleet is out on the fishing grounds near Denman Island. There are gulls everywhere. A sign that there are lots of fish around for the hungry gulls.

TIGHTER — Several seiners pull their nets through the water.

                    NARRATOR
          During the first 60 years of
          the last century . . .

FROM SHORE — Another collection of commercial fishing boats.

                    NARRATOR
       . . . intense fishing pressure
       caused major declines in wild
       fish populations.

EXT. COMMERCIAL FISHING BOATS — DAY

One boat is pulling in its seine net.

> NARRATOR
> Harvest reductions worked, and by the late 1970s wild fish populations were healthy and nearing historic levels of abundance.

FROM THE DECK — The seine net is pulled onto the deck filled with salmon.

> NARRATOR
> But something changed, and within a few years, scientists noticed something was wrong with the marine ecosystem along British Columbia's coast, now known as the Salish Sea.

EXT. BC COAST (MAP) — DAY

The scale includes the full range of the west coast of BC. The Salish Sea lights up.

DISSOLVE TO:

A BARGE — It can be seen through the window of a boat. It has some equipment for fish farms on the back.

ANOTHER TUG — It's pulling a single open net pen farm into place.

ALEXANDRA MORTON — with her young son.

They're in a Zodiac on the ocean. Alexandra points off camera at a fish farm.

INT. ALEXANDRA MORTON'S HOUSE — AFTERNOON

Establish. Interview. It's a cool, stormy day outside. Alexandra Morton is at her computer by the window.

TITLE: Alexandra Morton, Independent Biologist.

              ALEXANDRA MORTON
     Ah, one day a tugboat showed
     up with a little salmon farm
     behind it and ah, it's like
     "Oh, good idea," you know.
     It'll bring more people to my
     town.

EXT. ECHO BAY (2010) — DAY

The school looks closed and all is quiet.

ON THE SWING SET — The playground is empty. The school didn't survive.

              ALEXANDRA MORTON
     And we're always worried about
     the school staying open. So,
     we're hoping for more kids.

EXT. SHORELINE (LATE 1980s) — DAY

A shot over Alexandra's shoulder as she looks out into the ocean.

EXT. CHANNEL — DAY

An early fish farm, located in a quiet, peaceful bay.

TIGHTER — A worker tosses feed into the pen for the stock.

                    NARRATOR
        Salmon farming, it was thought,
        would offset the demand for
        wild fish and help preserve
        wild populations.

A worker is cleaning a deck while another tosses food into the pens.

RESUME — Morton interview.

                ALEXANDRA MORTON
        I liked the idea that there
        was going to be people
        living in various places out
        there because I'm running
        around . . .

EXT. OCEAN (20 YEARS AGO) — DAY

Alexandra is on a boat with her son, taking photos.

                    ALEXANDRA MORTON (O.S.)
               . . . in a little speedboat by
               myself or with a small child.
               And I liked the idea that
               there's other people out there
               in case I got into trouble. You
               know,

INT. ALEXANDRA MORTON'S HOUSE — AFTERNOON

She looks down at her computer screen.

                    ALEXANDRA MORTON
               . . . I saw them as shelter
               from the storm. Ah . . . but
               they turned out to be the
               storm.

EXT. OCEAN (1987) — DAY

The single fish farm pen sits in a pristine setting.

                    NARRATOR
               Alexandra Morton saw that fish
               farm in the fall of 1987.

PUSHING IN — The pen has a walkway around its perimeter.

                    NARRATOR
               But the first fish farm on
               the British Columbia coast was
               operational 15 years earlier,
               in 1972.

Early fish farms located on the BC coast were mom and pop operations. This farm, started in 1981 on Nelson Island by Brad and June Hope under the banner Tidal Rush Marine Farm, initially raised chinook, coho and steelhead. It became one of the first two operations allowed to import Atlantic salmon eggs into BC. (Photo credit: Brad Hope)

EXT. A MOM AND POP FISH FARM — DAY

It's tucked against a shoreline with a small shack on it.

ANOTHER ANGLE — panning the floating cages from shore.

EXT. ANOTHER SMALL FARM — DAY

It has about a dozen small net pens and a small shack.

>   CRAIG ORR (O.S.)
>   The history of salmon farming
>   is kind of peculiar in British
>   Columbia . . .

INT. CRAIG ORR'S OFFICE — DAY

Establish. Interview. It's a small office.

TITLE: Dr. Craig Orr, Executive Director, Watershed Watch Salmon Society.

>   CRAIG ORR
>   . . . uh. It started off as a
>   fairly low-key, ah, mom and
>   pop kind of operation along
>   the Sunshine Coast of British
>   Columbia.

EXT. NELSON ISLAND (1979) — DAY

A quiet bay is surrounded by a treed lot.

EXT. NELSON ISLAND FOREST ROAD — DAY

Brad and June Hope stroll down a dirt road near their property on the island.

>BRAD HOPE (O.S.)
>We live on the ocean. Does anybody raise salmon? And so we started looking around and there was a few licences . . .

INT. HOPE RANCH — DAY

Establish. Interview.

TITLE: Brad Hope, Former Owner, Tidal Rush Farms.

>BRAD HOPE
>. . . issued, I think we were either licence, I don't know whether we were either 4, 5 or 6, one of the early ones,

EXT. BC SUNSHINE COAST (MAP) — DAY

A dot shows the location of the Hope Farm among the first fish farms. The words "Tidal Rush Farm" fade up and out.

>BRAD HOPE
>. . . but there had been a person actually within 15 kilometres of us . . .

BACK TO MAP

Another dot lights up, and the words "First BC Farm, 1972" fade up.

              BRAD HOPE (O.S.)
    . . . who'd had licence number
    1. I think his name was Al
    Meneely. And he'd tried with
    Pacific salmon, and it hadn't
    gone anywhere.

                                FADE TO:

PHOTO — The Fisheries Research Board barge moored at the Pacific Biological Station.

ANOTHER PHOTO — The barge is now out on a remote research location.

                NARRATOR
    In the early 1970s, Canadian
    government scientists at
    the Fisheries Research
    Board began thinking about
    aquaculture . . .

DOCUMENT — FRONT PAGE — "A Brief on Mariculture."

MOVING DOWN — from the headline to the date, "1972."

HEADLINE — The *Fisherman*, "Trend is to Aquaculture," December 18, 1974.

By 1984, the provincial government had granted 15 licences for small farms like this one. The rush was on, and it wasn't long before hundreds of licences were granted across the BC coast. (Photo credit: June Hope)

The words "an abundance of as yet unpolluted fresh and marine waters" lift off the page.

> NARRATOR
> It was thought that Canada was well positioned to grow an aquaculture industry. And the primary reason was because Canada had "an abundance of as yet unpolluted fresh and marine waters."

EXT. COASTLINE (AERIAL MONTAGE) — DAY

Three exhilarating shots of the coastline demonstrate the vastness of the BC coast.

EXT. NANAIMO HARBOUR (AERIAL PHOTO) — DAY

Looking down on the Pacific Biological Station, located on the bank of the harbour. It has a small test farm pen attached to its wharf.

> BRAD HOPE (O.S.)
> Found out that there was a little . . .

RESUME — Brad Hope interview.

> BRAD HOPE
> . . . test project at the biological station in Nanaimo that had seen Norway, seen
> (MORE)

                    BRAD HOPE (CONT'D)
          Scotland and realized that
          there was a potential here and
          were really encouraging.

INT. PACIFIC BIOLOGICAL STATION PEN — DAY

Chum salmon thrash around inside the net pen.

EXT. OCEAN — DAY

Stormy seas rock a shoreline. A commercial fish boat bobs about in the waves near the shore.

TIGHTER

The fisherman pulls in his net.

                    NARRATOR
          And there were good reasons for
          government encouragement. In
          the early 1980s, Canada, like
          everyone else, was suffering
          through a tough recession.

GRAPH — An animated red line indicates the rising public debt and then rising interest rates through 1970-1990.

                    NARRATOR
          Public debt was growing
          rapidly, and more than doubled
          from under 91 billion in 1980
          to over 260 billion by 1984.

A YELLOW LINE — appears and indicates how interest rates rose from the 1970s to the early 1980s.

                    NARRATOR
        Escalating interest rates
        were compounding the debt and
        the federal government was
        desperate to boost revenue.

HEADLINE — The *Fisherman*, "93 Sites in Works — Gold Rush Mentality Hits Fish Farms," July 19, 1985.

                GREGORY MCDADE (O.S.)
        There also was this kind of
        gold rush mentality . . .

INT. GREGORY MCDADE'S BOARD ROOM — DAY

Establish. Interview. Gregory McDade represented Alexandra Morton in her court challenge on jurisdiction regarding fish farms.

Gregory sits in front of a bookcase.

TITLE: Gregory McDade, QC, Managing Partner, Ratcliffe and Company.

                    GREGORY MCDADE
        . . . that seems to occur early
        in any industry . . .

EXT. OCEAN — DAY

An early farm sits in the ocean. A farmer walks around the perimeter.

> GREGORY MCDADE (O.S.)
> . . . where a bunch of people in government think, isn't this industry the best thing since sliced bread?

EXT. EARLY FISH FARM — DAY

The farm Alexandra saw being towed earlier. It's now set up and operating.

EXT. TIDAL RUSH FARM — MORNING

Brad dips his net into one pen.

> GREGORY MCDADE (O.S.)
> Look at the way we can take these useless oceans and make a whole bunch of money from them . . .

IN A HATCHERY

Eggs are poured into a big trough for fertilization.

Buildings were added to provide shelter and storage for the staff running the farms. This farm, owned by the Gordon Group, was located near Port Hardy. (Photo credit: The Fisherman Publishing Society)

BC fish farm companies were beginning to consolidate by the mid-1990s and, as the companies grew, so did the size of the farms. This farm, located in Kyuquot Sound, was photographed in February 1997. (Photo credit: The Fisherman Publishing Society)

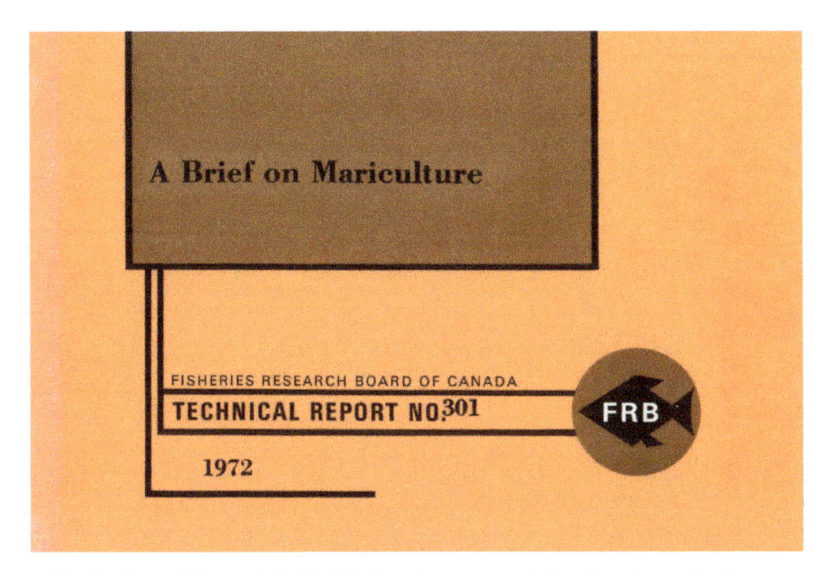

The findings of the study led by J.R. Brett in 1972 reveal that Canada was looking at developing an aquaculture industry. The Canadian government was looking to kick-start the industry, which was already growing in Norway and Japan. (Credit: Government of Canada)

In 1908, permanent biological stations were opened at St. Andrews, New Brunswick, and Nanaimo, British Columbia. In 1937, the Biological Board of Canada became the Fisheries Research Board of Canada, which conducted marine research until its demise in 1979. The biological stations are now part of the Department of Fisheries and Oceans, and scientists at Nanaimo experimented with raising chum salmon and other species in small net pens in Nanaimo Harbour next to the main research station. (Photo credit: The Fisherman Publishing Society)

>   GREGORY MCDADE (O.S.)
>   . . . by licensing them to
>   private corporations who will
>   then make money and pay us
>   taxes.

A HATCHERY WORKER

They're pouring fertilized eggs into an incubation tray.

>   GREGORY MCDADE (O.S.)
>   Governments, not only in Canada
>   but everywhere, seem to have a
>   great appetite for creating new
>   industry.

>   DISSOLVE TO:

HEADLINE — The *Fisherman*, "The politics of numbers," November 23, 1994.

HEADLINE — The *Fisherman*, "Whose Debt?" September 25, 1995.

>   NARRATOR
>   But boosting revenue was not
>   likely going to be enough to
>   balance the public books, so
>   politicians began to look at
>   spending.

INT. GEOFF MEGGS'S OFFICE — DAY

Establish. Interview. Geoff Meggs.

The Fisheries Research Board barge was used to conduct field marine research at different locations along BC's coast. (Photo credit: The Fisherman Publishing Society)

TITLE: Geoff Meggs, Former Editor, the *Fisherman*.

> GEOFF MEGGS
> I think that the government incentives were not debt-related so much as they were related to the cost of management of the wild stock fishery and to the cost of . . .

HEADLINE — The *Fisherman*, "UI targeting fish industry workers," January 19, 1994.

The words "commercial fishermen take $4.03 out of the UI fund for every dollar they contribute" jump off the page.

> GEOFF MEGGS (O.S.)
> . . . income assistance that was rendered to fishing communities, and those were the ones that we heard about all the time. You know, if you took the total . . .

RESUME — Meggs interview.

> GEOFF MEGGS
> . . . cost of the Department of Fisheries and Oceans in, ah, direct terms and then you put that against the, ah, landed
> (MORE)

> GEOFF MEGGS (CONT'D)
> value of the fish . . . It
> always seemed like a pretty
> good deal to me because you
> were spending, if that was what
> you were worried about . . .
> 250 million dollars a year or
> something and getting back
> about a billion dollars in
> landed value.

EXT. MARINA — DAY

Lots of commercial fishing boats tied up. Not working.

> NARRATOR
> But the federal government at
> the time did not think it was a
> good deal . . .

PHOTO — More fishing boats tied up.

> NARRATOR
> . . . and struck an inquiry led
> by Dr. Peter Pearse . . .

PANNING UP THE FRONT PAGE — "'TURNING THE TIDE: A New Policy For Canada's Pacific Fisheries' The Commission On Pacific Fisheries Policy: Final Report"

> NARRATOR
> . . . to examine Pacific
> Fisheries policy.

The Fisherman

The politics of numbers

The newspaper columns are starting to fill up with all the figures about the costs of unemployment insurance as Human Resource Development Lloyd Axworthy's travelling road show makes its way across the country, telling Canadians how they need to "reform" social security.

First, there is a comparison of industries, with the figures showing that workers in agriculture, forestry and fisheries take $4.03 out of the UI fund for every $1 they contribute. Then there is the figure from the study commissioned by Axworthy's department which shows that "frequent users" of UI — those who like shore plant workers, suffer frequent, regular layoffs — account for 38 per cent of all UI collected.

No one will dispute the numbers — if fact the inequity is part of the design of the UI system. It was intended to re-distribute income from those who are employed full time and year-round to those who suffer unemployment. It's also true that it enables workers with skills, like fishermen and shoreworkers, to stay with the industry and maintain the productivity which is so well known on this continent.

What Axworthy's media support won't show is the enormous cost to this country of unemployment and the devastation to communities and families that will result from the proposed cuts to UI for seasonal workers. They won't refer to the study that was prepared by two Quebec economists, Diane Bellemare and Lise Poulin Simon which showed that the direct cost to Canada of unemployment, in terms of lost revenue amounted to a staggering $109 billion in 1993. The two women showed that the cost to the government to create a job through direct public works is only two-thirds of the economic cost of allowing that person to remain unemployed.

The Liberal government was elected, as Jean Chretien himself put it, "with the number one objective of creating jobs to put Canadians back to work." But what do we have? One example is DFO program review, which will result in 1,200 departmental jobs being eliminated across the country — the people doing spawning assessments, the enforcement officers and statisticians. Licensing systems in the herring fishery are putting seine crew members on the beach. And now Ottawa is saying that the cost of giving them income support is too high.

Make no mistake: the cost of that policy direction is a downward spiral. No income equals no purchasing power equals layoffs elsewhere in the economy. And all the cuts in social programs won't fix the massive losses to government revenue that will result.

While we're on the subject of revenue, there's another set of figures you likely won't see in Axworthy's publicity package: the money that Ottawa gives up because of tax breaks to corporations. According to a study prepared by the Ontario Federation of Labour and the Ontario Coalition for Tax Justice, 134 profitable corporations paid no taxes at all in 1992. In fact, the National Union of Public and General Employees, in its submission to Finance Minister Paul Martin, reported that $8 billion could be raised by introducing a minimum tax of 20 per cent on the profits of corporations whose tax rate was only half as high as the official rate thanks to tax breaks.

And finally, there's Ottawa's policy of allowing companies to defer taxes — a policy which, incidentally, has allowed Alcan Aluminum to defer $955 million in taxes. Even if Alcan and others were to pay the going rate of interest on their deferred money, it would generate more than enough revenue to maintain our UI and other social programs intact. How about it, Mr. Martin?

DFO restructuring

SORRY ABOUT THAT: In the last issue of the Fisherman we erroneously reported that the Sunshine Coast was considered a separate area for unemployment purposes and only 13 weeks were required. Wrong. The mainland coast in that area is considered part of Vancouver island for UI purposes and the requirement is 14 weeks. One area we neglected to identify was southern B.C., which covers the upper Fraser Valley and southern interior. Fifteen weeks are required there to qualify.

•

NEED THAT HERRING! Some folks are dying to get our herring, and others find it something to live for. Enthusiastic consumers of fresh herring line up before daybreak to make sure they don't miss out on the best deal for the freshest fish in town at the UFAWU herring sales every year.

The UFAWU's Bruce Logan told us of one hardy senior citizen in his eighties who lined up at 6 a.m. Nov. 20 in the cold and wind for his herring. Standing in the lineup which stretched several hundred metres along the water front, the man collapsed. Bystanders helped him up, and he said he was okay and stayed in the line-up. Didn't want to miss out on that herring. Shortly afterwards he collapsed again, and somewhat shakily regained his feet, but wouldn't move out of the line to rest. He fell down a third time, suffering what appeared to be a heart attack. Logan reports that the man was carried to the CKNW broadcast trailer where CPR was applied until the ambulance arrived. Emergency service personnel had

FISH & SHIPS

to use electrical equipment to revive the victim and then transported him to hospital. A couple hours later he sent his son to buy his herring. Can't miss out on that herring.

•

TRAWLER'S MEETING: The UFAWU trawl committee met Nov. 22 to discuss the effect of the Labour Code on trawlers, IQs and the 1995 fishing plan. The committee set the date for the annual trawl meeting for Dec. 15, 9:30 a.m. at Maritime Labour Centre.

•

HALLOWE'EN DANCE: Local 2 member Jackie Campbell reports that the annual Hallowe'en Howwl was again a success with over 100 people attending.

Campbell said she would like say thank you to the following contributors for donations of prize packages:

• **LOVELY IRVIN FIGG**, shoreworkers' organizer by day, was one of the prize winners at the Hallowe'en Howwl.

Safeway at 1st and Renfrew, London Drugs on East Hastings, Gulf and Fraser Credit Union, Doug at Vick enterprises, Mike at Canadian Fishing Company, Terry Baird from Ocean's, Mike at the Princeton and Molson, Janet Duplisse and J.S. McMillan, and Lil and Gal Horback. Campbell would like to say a special thank you to young Ryan Leblanc, the only child who attended the pumpkin carving session for his exceptional efforts.

Irvin Figg won the mens costume prize for his revealing bunny suit, Lil Horback shared the best women's prize with a woman dressed as a male shoreworker, and Sue Hale and her partner Gary won the best couple's prize for their tap-dancing costumes.

•

CENTURY SAM: To celebrate the 100th anniversary of the Ministry of Agriculture, Fisheries and Food, the Ministry has adopted the theme of "A Century of Achievement" and is honouring the contributions of pioneers in the farming and fishing industries.

In order to qualify, a farm or fish processing company must have been active in its provincial fishing industry over the last century. Supporting documentation, including copies of personal fishing licences should accompany application forms. Completed application forms should be forwarded to the Ministry before Dec. 1, 1994.

For more information contact Tracey Michalski at the Ministry of Agriculture, Fisheries and Food in Victoria. The phone number is 356-2236.

Published by the Fisherman Publishing Society on the third Friday of each month. Second Class Mail. Publications Mail Registration No. 1576. Rate Code 3. Single copy $1, $20 per year, $30 foreign. Deadline Friday prior to publication. Member of CALM.

180 - 111 Victoria Drive, Vancouver, B.C. V5L 4C4
(604) 255-1366 Fax (604) 255-3162

Desktop Publishing by Angela Kenyon, OTEU 15

Sean Griffin — Editor
Michel Drouin — Asst. Editor
Dave Watt — Advertising

4 • THE FISHERMAN / NOVEMBER 23, 1994

Escalating interest rates in the 1980s pushed the Canadian government at the time to re-evaluate unemployment insurance benefits used by seasonal workers like commercial fishermen. This likely played a role in the drive for an aquaculture industry that could provide year-round employment. (Credit: The Fisherman Publishing Society)

TURNING THE TIDE
A New Policy For Canada's Pacific Fisheries

THE COMMISSION ON PACIFIC FISHERIES POLICY
FINAL REPORT

Peter H. Pearse, Commissioner
VANCOUVER - SEPTEMBER 1982

Dr. Peter H. Pearse was tasked with leading the inquiry "Turning The Tide: A New Policy For Canada's Pacific Fisheries." Among his recommendations, he encouraged the expansion of mariculture only if it would not disrupt the natural stocks or those who depend on them. (Credit: Government of Canada)

                    GEOFF MEGGS (O.S.)
               Resource economists like Pearse
               saw this differently. There was
               no royalty on the fish.

RESUME — Meggs interview.

                    GEOFF MEGGS
               So, if you extracted a tree
               you paid stumpage, if you
               extracted a ton of ore you paid
               a royalty, but if you extracted
               a fish you didn't pay anything
               to the government.

HEADLINE — The *Fisherman*, "Privatization, Pearse still on DFO agenda," June 21, 1985.

HEADLINE — The *Fisherman*, "1984 salmon regulations — DFO tightens the economic screws," April 19, 1984.

HEADLINE — The *Fisherman*, "Tories hire 700 to hit UI users," December 12, 1984.

                    GEOGG MEGGS (O.S.)
               They needed privatization on
               that front . . .

HEADLINE — The *Fisherman*, "UFAWU claims victory on UI changes," May 22, 1987.

RESUME — Meggs interview.

>   GEOFF MEGGS
> . . . but also eliminate the
> people who were claiming
> unemployment insurance and
> so on, because they would no
> longer have an industry to
> participate in.

EXT. UCLUELET WHARF — DAY

Establish. Interview. Commercial fisherman Allan Gudbranson stands on a wharf.

TITLE: Allan Gudbranson, Commercial Fisherman.

>   ALLAN GUDBRANSON
> The Royal Bank was going to
> lend me money in 1992, '93,
> somewhere in there, and when I
> phoned him back the next year,
> he said no chance of getting
> any money. And I said why, and
> he said because BC Packers told
> us not to lend you fishermen
> any money.

Camera person Anissa Reed asks questions off camera.

>   ANISSA REED (O.S.)
> Why?

                    ALLAN GUDBRANSON
          Because we were going to be
          put out of business. And so
          the very next year, when that
          was happening, Fisheries was
          putting bulletproof glass
          in their Fisheries offices,
          and they cancelled all the
          fishermen's insurance, because
          they knew what they were going
          to do. They knew what they were
          going to do . . .

                    ANISSA REED (O.S.)
          What were they going to do?

                    ALLAN GUDBRANSON
          Screw the west coast fishermen,
          the small guys.

                                        FADE OUT:

A few seconds of black.

FADE IN:

HEADLINE — The *Fisherman*, "Aquaculture:
Fledgling industry offer perils and promise,"
April 19, 1984.

                    PETER PEARSE (O.S.)
          Fish farming was something
          that was hardly developed at
          all when I got going on my
          commission . . .

PHOTO — An early octagonal-shaped fish farm in a bay.

                PETER PEARSE (O.S.)
    . . . but it seemed to me that
    there was quite an opportunity
    here . . .

INT. PETER PEARSE'S HOME — DAY

Establish. Interview. He's sitting in his home office.

TITLE: Dr. Peter Pearse, Professor Emeritus, the University of British Columbia.

                PETER PEARSE
    . . . and it was only going to
    be a matter of time before we
    would have an industry based on
    fish farming.

PHOTO — A slightly larger farm with red nets fills the frame.

ANOTHER PHOTO — The same farm.

                PETER PEARSE (O.S.)
    But I think I did portray a
    cautious approach to fish
    farming to accommodate it.

BACK TO DOCUMENT — "Turning the Tide — A New Policy For Canada's Pacific Fisheries." September 1982.

The words "policies for accommodating new ventures" and "natural fish stocks or those who depend on them" lift off the page.

> NARRATOR
> Pearse did recommend that any "policies for accommodating new ventures" should not disrupt "natural fish stocks or those who depend on them."

EXT. BC SOUTH COAST (MAP) — DAY

Eighty tenures are marked on the map, with dots springing up one after the other, like bullets hitting a target.

> NARRATOR
> But Pearse's words were lost in the stampede by prospective fish farmers staking tenures and the provincial government scrambling to accommodate them.

FADE TO:

EXT. ST. ANDREWS — NEW BRUNSWICK (1983) — DAY

Establish. Aerial photo of the Biological Station.

                    NARRATOR
          In July 1983, the Canadian
          government held a conference
          at St. Andrews, New Brunswick,
          with the goal of organizing the
          new aquaculture industry.

DOCUMENT — "Aquaculture: A Development Plan for Canada"

FINDING THE WORDS — "Final Report of the Industry Task Force on Aquaculture Sponsored by"

                    NARRATOR
          The first step was the
          formation of a task force . . .

PAN DOWN — to the words "Science Council of Canada."

                    NARRATOR
          . . . that would be assisted by
          the Science Council of Canada.

RESUME — Meggs interview.

                    GEOFF MEGGS
          The Science Council was
          promoting salmon farming and
          was intervening in the industry
          to help it overcome these
          technological problems,
                    (MORE)

                    GEOFF MEGGS (CONT'D)
               not to highlight them or to,
               ah, or to suggest there was any
               problem with salmon farming.

EXT. OCEAN (AERIAL) — DAY

Two fish farms in the ocean.

                    NARRATOR
               In fact, the task force
               was less about science and
               technology.

DOCUMENT — "Aquaculture: A Development Plan for Canada."

Moving down the page showing the organizations and members of the task force.

                    NARRATOR
               It was more about organizing
               business interests. And these
               business interests declared
               their intentions the following
               spring, when at the next . . .

HEADLINE — The *Fisherman*, "Aquaculture: Fledgling industry offers peril and promise," April 19, 1984.

                    NARRATOR
          . . . conference in Vancouver
          it was boldly announced
          that . . .

THE WORDS — "common property fishing are over" lift off the page.

                    NARRATOR
          . . . the days of "common
          property fishing are over."

RESUME — Meggs interview.

                    GEOFF MEGGS
          Salmon and wild resources
          generally, in my view, are
          best protected in what I would
          call a common property setting
          where everybody has a stake
          in the outcome. It's not an
          unregulated setting, and so
          Garrett Hardin, who famously
          coined the Tragedy of the
          Commons, the idea that, ah,
          in the absence of private
          property rights there would
          be an extinction of natural
          resources, was very much the
          ideologue who drove fisheries
          policy, and still does drive
          fisheries policy, in this
          country and many others . . .

Aquaculture:
A Development Plan for Canada

Final report of the
Industry Task Force on
Aquaculture sponsored by
the Science Council of Canada

SH
37
S352
1984
UAFM

A task force called the Science Council of Canada was made up of early fish farm owners who completed their final report in August 1984. The report outlined the opportunities and needs facing the new industry. It was less about the science, though, and more about the business case for the new industry. (Credit: Government of Canada)

AQUACULTURE
Fledgling industry offers peril and promise

By GEOFF MEGGS

AQUACULTURE, the fledgling technology that Peter Pearse believed could be combined with ocean ranching to ease the problems of the fishing industry, is gaining momentum in B.C.

More than 200 biologists, civil servants, fish marketers and hopeful entrepreneurs gathered in a penthouse ballroom high above Vancouver last month to see the Science Council of Canada unveil a discussion paper designed to lay the groundwork for a Canadian aquaculture industrial development plan.

"We must act now or lose a commercial and employment opportunity," task force member David Saxby told the gathering. Science Council representative Ann Levi-Lloyd urged all present to submit their views quickly, "because we want to get this industry launched as soon as possible."

"The days of common property fishing are over," says one Science Council release. "Given the finite ability of the ocean to produce fish — especially in its polluted state — and the steadily rising costs of hunting the wild schools, as well as the urgent need to restructure our fishing industry, aquaculture (is) a compelling alternative.

"A doubling of present aquaculture yields is likely to create one million new jobs world-wide."

Jobs, especially for displaced fishermen, easier fisheries management, and increased food production are the three main advantages aquaculture promoters claim for their industry.

But those directly involved in aquaculture, particularly the fish farmers themselves, are much more conservative. Those with first-hand experience in Europe or British Columbia warn that aquaculture will not develop in this country without a major government commitment, including new legislation, financial assistance, research and development aid, and marketing analysis.

Even if this support is forthcoming, it will be five and more likely 10 years before aquaculture production takes off, they say, and employment benefits will not be enough to substitute for declines in the traditional fishery.

Aquaculture is the rearing of marine plants and animals in salt or fresh water. It includes everything from the propagation of seaweed, a substantial industry in Japan, to the rearing of oysters and the farming of salmon. Ocean ranching, the private-for-profit production of salmon for release to the wild, is included in most discussions of aquaculture.

Although aquaculture of all products will hit a worldwide total of 10 million tonnes in 1984, about 10 per cent of the planet's harvest of marine products, Canada produced a mere 2,850 tonnes in 1982. About 300 tonnes of that amount was salmon, compared to Norway's 15,000 tonnes.

Yet Canada's long and relatively clean coastline make it one of the world's most promising nations for increased aquacultural production, particularly in valued species like salmon.

With the prompting of the Science Council and the support of key individuals in private industry and government, advocates of aquaculture have been gaining ground. A national conference in New Brunswick in 1983 produced the task force and it has produced its draft development policy.

Aquaculture is on the agenda in Canada and the debate about who might benefit from its production has begun.

Norway, the acknowledged giant of salmon farming, expects to hit production of 30,000 tonnes in the next few years. But the Norwegian experience, while impressive, belies claims that salmon farming can be a decentralized, job producing industry characterized by small business ownership.

A/S Mowi is Norway's dominant farmed salmon producer. Company president Thor Mowinckel told delegates to the World Mariculture Society conference in Vancouver last month that it took 14 years and millions of dollars for farming to become profitable after experiments began in 1981.

Mowi now has expanded to Iceland, Ireland and Scotland and is experimenting with halibut and sole, but its success was built on the strong financial backing it received from Norsk Hydro, the country's major utility.

Although 490 farms were operating in 1983, Mowinckel said, most relied on Mowi for research, smolts, feed, marketing and even financing. "We are establishing vertical integration," he said, "and I strongly recommend co-operation between large and small companies."

Equally revealing was his disclosure that the industry directly employs some 1,500 workers this year, a number that will grow slowly as expansion continues. By eliminating the fishing fleet and reducing processing to a simple gutting process, employment is dramatically reduced.

With the task force estimating that only one person-year of employment is generated for every 20 tonnes of production, it is clear that salmon farming of 30,000 tonnes of salmon would employ only a fraction of those now employed in the industry.

With government price support and a district currency advantage, Mowi is reaping good profits at last, thanks to the high quality of its production, the premium restaurant market it commands and the year-round availability of its product.

Given the success in Norway, what is holding up Canada? According to aquaculturalists, just about everything. The Science Council Task Force, which included Powell River salmon farmer Brad Hope, is demanding the development of a national aquaculture policy.

In the short term, bolstered by immediate increases in funding from the fisheries department budget, the industry would sit down with related government agencies in provincial aquaculture co-ordination committees to push for immediate development of species like oysters, salmon, roe on kelp and trout.

The department of fisheries would be charged with the creation of a national secretariat to fund aquaculture development and encourage the creation of a national policy. New legislation would be drafted to protect and regulate the new industry and research funding would be directed to the needs of aquaculture.

So far, no "traditional" fisheries groups have been involved in the growing discussion of aquaculture, although B.C. Packers has long maintained an interest in the area and was represented at the conference. In fact, some aquaculturalists see "traditional" dependance on conventional fisheries" as an obstacle to their plans for growth and development.

Others, like Hope, believe the existing fisheries sector can be an important ally. "People are coming to us constantly and saying they are fishermen and they see the writing on the wall and they want to get into farming," he says.

He believes that a buyback could give some the money to move into farming, but he warns at least $300,000 must be available to get started.

And although he is aware of the resistance to ocean ranching in the industry, he insists that "done properly and not on the Alaska model it provides great potential to put more out there to be caught."

He fears, however, that the department of fisheries will see farming as a substitute for existing commercial fisheries and "say the commercial fishery is a thing of the past."

As president of the B.C. Mariculture Association, Hope is lobbying for major government initiatives to give the fledgling industry a sound start.

A major requirement is legislation to establish property rights in the fish produced, to establish standards for disease control, sanitation and management and to ensure provision of aquacultural sites with high water quality.

Further research is needed, largely by the public sector, into genetic development and nutrition. Financing assistance similar to that provided to land farmers also will be required to enable banks to lend money to entrepreneurs.

Finally, aquaculturalists hope the government will help develop a marketing strategy to give them better access to areas already served by the aggressive Norwegians.

But the exports at the Mariculture Society convention agreed the Canadian industry faces more years in the wilderness before it can offer anything to the economy or the industry.

European experience indicates the industry will go through a period of constant business failures before it stabilizes, warns Professor J. A. Spence. This pioneer phase could last 10 years and has really just begun.

British Columbia, unlike Europe has a special problem relatively strong wild stocks supporting a large conventional industry.

"If you have a large farmed salmon industry it could potentially eat into wild salmon markets," he warned. Social and economic impacts could be severe.

Art McKay, a Nova Scotia salmon farmer, was more blunt. Although enthusiastic about the industry's prospects, growth entails "enormous capital costs and the risks are very, very high."

Next issue: Brad Hope's Total Rush Marine Farms. Are his 14-pound brood stock chinook the wave of the future?

THE FISHERMAN — APRIL 19, 1984/7

Geoff Meggs, editor of the *Fisherman* newspaper at the time, reported that a member of the Science Council task force announced at a conference that the days of "common property fishing are over." (Credit: The Fisherman Publishing Society)

EXT. COMMERCIAL FISHING BOAT — DAY

A seine net full of salmon is pulled onto the deck.

> GEOFF MEGGS (O.S.)
> . . . and in the case of the BC coast, there was a tremendous battle over privatization and the extinction of common property fisheries.

HEADLINE — The *Fisherman*, "Alcan seeks Kemano approval," January 20, 1984.

> NARRATOR
> The battle between common property fisheries . . .

EXT. RIVER — DAY

Sockeye salmon are at the spawning grounds. Their red backs are sticking out of the water.

> NARRATOR
> . . . and aquaculture took on added complexities when it appeared other business interests . . .

HEADLINE — The *Fisherman*, "Moratorium threatened: Green light given for offshore drilling," May 21, 1986.

                    NARRATOR
          . . . might support the end of
          wild fish stocks.

EXT. NORWEGIAN BUILDING — DAY

The Norwegian flag flies in the foreground.

EXT. NORWEGIAN FJORD — DAY

Establish. An expansive view of the ocean.

EXT. NORWEGIAN RIVER — DAY

A small waterfall plunges over the rocks.

TIGHTER ON THE WATER

A single Atlantic salmon leaps into the air. Then another.

                    GEOFF MEGGS (O.S.)
          We discovered that in
          Norway . . .

EXT. NORWEGIAN FISH FARM — DAY

Snow-capped mountains fill the background.

                    GEOFF MEGGS (O.S.)
          . . . salmon farming was
          designed to replace wild
          fish . . .

EXT. ANOTHER NORWEGIAN FISH FARM — DAY

Another fish farm in the ocean off the Norwegian coast.

> GEOFF MEGGS (O.S.)
> . . . so that rivers and watersheds would be available for other purposes. So we saw a, ah . . .

RESUME — Meggs interview.

> GEOFF MEGGS
> . . . a country that was farming because it had wiped out its wild stocks.

EXT. BC FISH FARM — DAY

It lies tight along a shoreline.

> NARRATOR
> The idea that other industries might be eyeing . . .

EXT. SHORELINE — NIGHT

A stormy channel.

> NARRATOR
> . . . the spaces occupied by wild fish sent shivers through British Columbia communities.

ANOTHER SHOT

An empty ferry dock lies between the shoreline and a small BC community.

EXT. SALISH SEA — DAY

A misty day. Two fish boats work the waters for fish.

                    NARRATOR
          But changing Canada's west
          coast culture from a wild fish
          economy to . . .

EXT. NARROW CHANNEL — DAY

A fish farm with round pens occupies a stretch along the shoreline.

                    NARRATOR
          . . . a private aquacultural
          one would not be easy.

                                        FADE TO:

EXT. TREE-COVERED POINT — DAY

Mist flows on water in the distance.

EXT. SOINTULA WHARF — DAY

Alexandra Morton looks out at the stormy weather.

56.

RESUME — Morton interview.

              ALEXANDRA MORTON
    DFO did a fast one. There
    used to be what was called the
    Fisheries Research Board.

MONTAGE — Technical reports produced by the Fisheries Research Board of Canada.

              ALEXANDRA MORTON (O.S.)
    [It] produced a lot of the
    groundbreaking science on
    different species of fish.
    They took that Fisheries
    Research Board . . .

RESUME — Morton interview.

              ALEXANDRA MORTON
    . . . and they pulled it into
    the political body of DFO and
    they renamed it the Pacific
    Biological Station. And
    suddenly science was dictated
    by politics.

EXT. PACIFIC BIOLOGICAL STATION — DAY

Establish. An old sign: "Pacific Biological Station."

FROM ACROSS THE BAY

DFO's research vessel is docked next to the building.

> NARRATOR
> The elimination of the Fisheries Research Board perhaps marked the beginning of a significant shift in the way Canada's civil service would work.

INT. PODIUM (CBC) — DAY

Brian Mulroney at a news conference.

> OTTO LANGER (O.S.)
> Our minister came right here from Richmond. I remember, ah, prior to the Mulroney government,

INT. OTTO LANGER'S HOME — DAY

Establish. Interview.

TITLE: Otto Langer, Retired DFO Fisheries Biologist.

> OTTO LANGER
> . . . some of us didn't see that much political interference. And then when the Mulroney government came into power, it was an overly politicized system.

EXT. TENT — DAY

Prime Minister Mulroney is talking with someone off camera.

>   JOHN VOLPE (O.S.)
>   I made the mistake of assuming
>   that DFO . . .

INT. VOLPE'S OFFICE — DAY

Establish. Interview.

TITLE: Principal Researcher, SEG Group, University of Victoria.

>   JOHN VOLPE
>   . . . was interested in
>   managing the resource. Right.
>   Um, finally, you know, the
>   eureka moment.

INT. NEWS CONFERENCE — DAY

Dave Gillespie, commissioner of the *Gillespie Report*, is addressing the media.

>   JOHN VOLPE (O.S.)
>   No. DFO no longer manages
>   resources, they manage issues.

AT A PODIUM — Gillespie addresses a crowd.

                    DAVE GILLESPIE
          I am satisfied that the, um,
          that the, ah, the fish farms,
          that are in existence and the
          fish farms that are now being
          applied for are not going to
          be a serious, ah, environmental
          hazard on the coast of British
          Columbia.

RESUME — Langer interview.

                    OTTO LANGER
          When people do come back after
          two years in Ottawa, they're
          almost a different animal. They
          even talk differently and they
          think differently and they're
          totally brainwashed into . . .

INT. NEWS CONFERENCE — DAY

The Federal Fisheries Minister, Tom Siddon, is at the podium.

                    OTTO LANGER (O.S.)
          . . . you've got to protect
          the minister at all costs. The
          minister is never wrong.

INT. NEWS CONFERENCE — MOMENTS LATER

The Provincial Minister of Agriculture and Fisheries, Brian Savage, is now speaking.

OTTO LANGER (O.S.)
And you look at this fellow, and a year or two ago he was your friend and he shared your values. He comes back from Ottawa . . .

RESUME — Langer interview.

OTTO LANGER
. . . and you almost don't understand his thinking processes any more, and he's lost sight of fish. Everything's about briefing notes, keeping Ottawa informed, not getting the minister upset, don't create any waves, and you become the director general of noise control.

EXT. NEWFOUNDLAND DOCK (FLASHBACK) — DAY

A large gathering of fishermen holding cardboard signs.

NARRATOR
And by the mid-1980s, the noise was getting louder on both coasts as fishermen were riled by what appeared to be gross mismanagement of Canada's fisheries.

EXT. NEWFOUNDLAND DOCK - MOMENTS LATER

John Crosbie, then the Fisheries Minister, is in the middle of an angry crowd of fishermen.

                   JOHN CROSBIE
      There is no need to abuse me.

                   ANGRY FISHERMAN
      I'm not abusing you! I didn't
      abuse you.

                   JOHN CROSBIE
      I didn't take the fish from the
      God-damned water.

                   ANGRY FISHERMAN
      Then who took it then?

                   JOHN CROSBIE
      Don't go abusing me.

INT. JOHN CUMMINS OFFICE - DAY

Establish. Interview.

TITLE: John Cummins, Former Canadian Member of Parliament.

                   JOHN CUMMINS
      The department took over in
      1985, and it was a complete
      flip-flop. All of a sudden they
      got into, ah, to a weak stock
                      (MORE)

JOHN CUMMINS (CONT'D)
management, they reduced fleet size, they started plugging the spawning grounds with . . .

HEADLINE — The *Fisherman*, "War on fishermen: Total closure on Fraser," January 30, 1981.

JOHN CUMMINS (O.S.)
. . . with fish. But it's their management of the fishery, that's where the downfall's happened.

PHOTO — John Fraser in a media scrum.

JOHN FRASER (O.S.)
Well, I became Fisheries Minister following the election in 1984,

INT. LIVING ROOM — DAY

Establish. Interview.

TITLE: John Fraser, QC, Former Fisheries Minister and House Speaker.

JOHN FRASER
. . . and some of the people who came to see me were from the Maritime provinces.

EXT. DOCKS — DAY

Commercial fishing boats are tied up at the docks.

> JOHN FRASER (O.S.)
> And what they said to me was that if we continue the commercial netting of Atlantic salmon, we're probably going to run the salmon populations down so low that for all practical purposes they will be . . .

RESUME — Fraser interview.

> JOHN FRASER
> . . . extirpated. Which means that there's no commercial value left at all.

EXT. COMMERCIAL FISHING VESSELS — DAY

East coast fishing vessels are tied up at the docks. Idle.

> JOHN FRASER (O.S.)
> Before I became minister, I said publicly,

HEADLINE — The *Fisherman*, "Applauds Survival Coalition: Fraser stands by election pledges," January 18, 1985.

> JOHN FRASER (O.S.)
> . . . "We are going to enter into an arrangement to buy back the licences."

EXT. EAST COAST COMMUNITY — DAY

Establish. A peaceful setting.

> JOHN FRASER (O.S.)
> In other words, we wouldn't confiscate, we . . . we'd expropriate, in effect . . .

EXT. MARINA — DAY

Fishing boats sit idle.

> JOHN FRASER (O.S.)
> . . . which was fair. The first few years after we stopped the netting . . .

RESUME — Fraser interview.

> JOHN FRASER
> . . . the number of large salmon coming back increased quite dramatically for some years. Now, they got some ocean problems much later but that's another story.

Word leaked out and was reported in the *Fisherman* that the Department of Fisheries and Oceans was planning to sharply curtail Fraser River commercial fisheries because many salmon populations were declining rapidly. No one knew what was causing the declines, but commercial fishing became the first target. (Credit: The Fisherman Publishing Society)

EXT. ATLANTIC OCEAN — DAY

A seine net is being drawn over the side of a boat. A lone Atlantic salmon is pulled on board.

> NARRATOR
> But was it another story? Or did the closing of the Atlantic salmon fishery just buy a bit of time before the decline continued?

A FISHERMAN — He untangles the salmon from his net.

> NARRATOR
> And then came the closure of the Northern cod fishery.

INT. HALLWAY (CBC) — DAY

Fishermen try to break down a door.

> REPORTER
> Fishermen stormed the doors of John Crosbie's news conference.

Two men try to shoulder the door open.

> REPORTER
> They were enraged over a compensation package they say is unacceptable.

Fish populations began to crash on Canada's east coast in the mid-1980s, and the Canadian government at the time closed the Atlantic salmon fishery. By 1992, Northern cod (*Gadus morhua*) populations had declined to 1% of their former abundance, forcing the closure of one of the greatest fisheries in the world. (Photo credit: NOAA)

68.

INT. CONFERENCE ROOM HALL (CBC ARCHIVE) — DAY

Four security guards grab a chair and use it to jam the doors to prevent the fishermen from entering the room.

> REPORTER
> Security guards locked the doors and frantically called for help.

TIGHT ON GUARD

> SECURITY GUARD
> Call the police for back-up.

INT. CONFERENCE ROOM (CBC ARCHIVE) — DAY

The room is filled with media, and Fisheries Minister John Crosbie sits at the front of the room, ready to make his announcement.

> REPORTER
> Inside, the Fisheries Minister tried to ignore the banging . . .

FROM THE MINISTER'S RIGHT

> REPORTER
> . . . and explained why he just shut down a 400-year-old fishery.

                    JOHN CROSBIE
          I'm making a decision based
          on the desire to ensure that
          the Northern cod survives as a
          species.

INT. OCEAN — DAY

Cod swim slowly as they forage.

                    NARRATOR
          By the early 1990s, the North
          Atlantic cod population had
          plunged to approximately one
          percent of its normal level of
          abundance.

GRAPH — It shows the Northern cod catch trend over a few decades and then the sudden drop.

BACK TO THE COD — It's an impressive school.

INT. DANIEL PAULY'S OFFICE — DAY

Establish. Interview.

TITLE: Dr. Daniel Pauly, Professor, Fisheries Centre, the University of British Columbia.

                    DANIEL PAULY (O.S.)
          When I was a student, Canada
          was in high esteem, was
          viewed in high esteem. And the
          Canadian journal was, you know,
          like a bible.

INT. COMMERCIAL FISH BOAT — DAY

A fisherman kicks cod into the hold of a boat. Another shovels the mixed catch onto a conveyor belt.

> DANIEL PAULY (O.S.)
> And ah . . . and all of this was very much questioned when the stocks of Northern cod collapsed because there was . . .

INT. OCEAN — DAY

Dead cod lie on the ocean floor.

> DANIEL PAULY (O.S.)
> . . . there was the Canadian prime stock.

INT. DANIEL PAULY'S OFFICE — DAY

RESUME — Pauly interview.

> DANIEL PAULY
> The, the, the stuff that, that made Canada, ah, that went down the tube . . . people blaming each other and so on.

INT. OCEAN — DAY

Another shot of dead cod on the ocean floor.

                    DANIEL PAULY
                And so what happened?

RESUME — NEWFOUNDLAND DOCK

The fight is still on.

                    JOHN CROSBIE
                Don't go abusing me!

                    ANGRY FISHERMAN
                You and your people took it!

                    JOHN CROSBIE
                Now, I'm trying to do what I
                can to help.

                    ANGRY FISHERMAN
                You're doing shit all. You're
                doing nudding.

Crosbie tries to get through the crowd.

                    ANGRY FISHERMAN
                You're doin' nudding. You're
                doin' nuddin'.

HEADLINE — The *Fisherman*, "Comeback on the cod: Newfoundland's 'sentinel fishermen' tracking the revival of the cod stocks," October 21, 1996.

                    JOHN CUMMINS (O.S.)
               Behind it all, the department
               was saying, well, you know
               we've got this moratorium on
               and, you know . . .

RESUME — Cummins interview.

                    JOHN CUMMINS
               . . . in four or five years
               things are going to be back to
               normal. Stocks are going to
               recover, everything's going to
               be fine. Well, you know, here
               we are almost 20 years later,
               and they're not back. The
               stocks didn't recover.

ATLANTIC WILD FISH GRAPH — Lines trace
the abundance history of three wild fish
populations over time. First Atlantic salmon,
then cod, then capelin.

                    NARRATOR
               But this wasn't just an
               Atlantic salmon story, or a
               cod story. Many species of
               fish were crashing on the east
               coast. And the same thing was
               happening to Pacific salmon on
               the west coast.

PACIFIC WILD FISH GRAPH — This time Pacific
salmon abundance is traced by species, and

all the populations take a dive in the early
1990s.

EXT. SEINE NET — DAY

It's full of salmon. It's a massive catch.

WIDER — A fisherman has a wriggling mass of
fish on his deck.

> NARRATOR
> Was it overfishing, climate
> change, poaching? Everyone
> was looking for someone or
> something to blame.

                                       FADE OUT:

FADE IN:

EXT. SOINTULA FERRY — DAY

Alexandra Morton and several supporters
arrive at the ferry. They have signs in their
hands.

> CROWD
> Go, Alex, go! Go, Alex, go! Go,
> Alex, go!

Alexandra walks onto the ferry with her dog
and with a flag sticking out of her backpack.

**SUPERSCRIPT: Get Out Migration Protest —
Day 1**

Alexandra Morton leaves Port McNeill at the start of the Get Out Migration, which culminated 16 days later at the BC legislature. Approximately 5,000 people were in attendance. (Photo credit: John Preston)

By the time the protest reached Fanny Bay, partway down Vancouver Island, the number of protesters had grown substantially. (Photo credit: John Preston)

ON THE SOINTULA FERRY

The ferry pulls away from the dock.

MOMENTS LATER

The ferry is now in the middle of the channel. Alexandra is alone and taking a photo with her cell phone.

TIGHT ON A SIGN — The sign reads "Salmon Are Sacred."

EXT. BEACH — DAY

Someone has scrawled in the sand the words "The Tide is Turning." A wave washes over the words.

					BACK TO:

RESUME — Morton interview.

		ALEXANDRA MORTON
	It has to be, you know, what
	Gandhi did. Peaceful, but
	resolute, decision. You are not
	gonna take our salmon away.

EXT. PORT MCNEILL STREET — DAY

Alexandra leads the demonstration off the ferry.

                    ALEXANDRA MORTON
               Hey, Howard. Good to see you.

EXT. PORT MCNEILL STREET — DAY

The demonstration climbs the hill, exiting
Port McNeill.

TIGHTER

Alexandra is in the lead. One demonstrator
holds a sign saying "No Salmon No Future."

EXT. ISLAND HIGHWAY — DAY

The protesters have now walked several
kilometres, some of them down the middle of
the highway.

SOME TIME LATER

They cross a bridge and fill the highway.

EXT. DIRT ROAD — DAY

The protesters make their way toward a
hatchery for a rally.

                                       FADE OUT:

FADE IN:

EXT. RIVER — DAY

Low angle looking across the water. The back of a sockeye salmon barely comes out of the water.

LOOKING DOWN — In the water we see quite a few sockeye salmon trying to spawn.

INT. RIVER — DAY

A large school of red sockeye salmon mill around in a deep pool.

> NARRATOR
> When the federal government was . . .

EXT. SOINTULA — EVENING

Across the channel, the small town of Alert Bay is nestled along the stormy shoreline. Salmon leap into the air.

> NARRATOR
> . . . shutting down the wild salmon fisheries that were so important to British Columbia's coastal communities . . .

EXT. FISH FARM — DAY

A grid of net pens stretch toward shore.

                    NARRATOR
          . . . the provincial
          government's response was
          to . . .

TIGHTER — A man looks over his pens of fish.

                    NARRATOR
          . . . ramp up aquaculture even
          more as a replacement.

                                        DISSOLVE TO:

INT. TIDAL RUSH FARM (1980s PROVINCE OF BC
COMMERCIAL) — DAY

A young Brad Hope appears in a government
commercial.

                    BRAD HOPE
          We're using the most advanced
          mariculture technologies
          available to us in the world
          today. The Europeans have a 20-
          year head start on us in this
          industry. We're going to catch
          up quickly and compete with
          them.

ON A MICROSCOPE — Brad Hope leans in and
looks into the eyepiece.

                    COMMERCIAL NARRATOR (V.O.)
          New products are creating new
          jobs.

79.

UNDERWATER — Atlantic salmon swim around in a net pen.

EXT. OPEN NET PEN — DAY

June Hope tosses a scoop of fish food into a pen.

INT. FISH FARM LAB — DAY

A lab technician takes a sample from a salmon.

> BOB MCCLELLAND (O.S.)
> We have to find and develop new ideas, new products . . .

EXT. WHARF — DAY

June Hope is examining eggs in a tray.

INT. BOB MCCLELLAND'S OFFICE (1980s) — DAY

Establish. Interview. He is sitting at his desk.

TITLE: Bob McClelland, Minister of Industry.

> BOB MCCLELLAND
> . . . and new markets. Our job as government is to provide support and encouragement that will turn these new opportunities into new jobs.

INSIDE A PEN — Atlantic salmon swim around in a dense school.

SUPERSCRIPT: Partners in Enterprise, Parliament Buildings, Victoria, V8V 1X4. Province of British Columbia.

EXT. SECHELT FARM (1980s) — DAY

A few workers are working a small farm.

> NARRATOR
> Aquaculture was so inviting to the provincial government . . .

MONTAGE — Aquaculture licences land one after the other on a table. Slap, slap, slap. With the last one we hear the thump of a stamp to emphasize the "stamp of 'Approval.'"

> NARRATOR
> . . . that by the spring of 1986, it was approving more than one licence per day for months.

EXT. SMALL FISH FARM — DAY

Establish.

> GEOFF MEGGS (O.S.)
> The industry was, it was treated like a mining claim, ah, where you, like the oil
> (MORE)

                    GEOFF MEGGS (O.S.) (CONT'D)
                    and gas industry, you may have
                    a farm, you can negotiate how
                    access is achieved. But you
                    can't deny . . .

RESUME — Meggs interview.

                    GEOFF MEGGS
                    . . . access to a drilling
                    company in Alberta, and it felt
                    the same way to me in terms of
                    access to these leases.

BACK TO THE SMALL FARM — It's located near a gravel conveyor belt.

                    GEOFF MEGGS (O.S.)
                    You know, there were certain
                    locations because of their
                    exposure to wind and tide and
                    so on that were better suited
                    than others.

PHOTO — Another farm is located near homes.

                    GEOFF MEGGS (O.S.)
                    And so they would be staked,
                    and, and often they were staked
                    in a pre-emptive way to stop
                    others from having them.

                                        FADE TO:

EXT. BROOD STOCK POND — DAY

A worker at Tidal Rush farms grabs a chinook salmon by the tail and lifts it out of a net. June Hope takes a look at it.

> BRAD HOPE (O.S.)
> At the beginning, I guess we didn't realize that, as they do in Scotland and Norway, they moved fish a long ways, so they don't seem to care how far away from . . .

RESUME — Hope interview.

> BRAD HOPE
> . . . the market the fish is. They do large numbers and they get them there economically. We were trying to be closer to market . . .

PHOTO — Workers lift a plankton pen out of the water.

> BRAD HOPE (O.S.)
> . . . and where you could get staff and staff could get in and out.

BACK TO THE POND — A mature chinook is lifted into a partitioned area of the holding pond.

                    BRAD HOPE (O.S.)
          So we really looked at a factor
          that shouldn't have come into
          play.

INT. CHINOOK FISH FARM (MID-1980s) — DAY

Chinook salmon are jammed together in a pen.

SUPERSCRIPT: Chinook fish farm mid-1980s

                    BRAD HOPE (O.S.)
          You need the environment for
          the fish.

                    NARRATOR
          Locating farms close to
          infrastructure made sense from
          an operations standpoint. But
          the communities they were next
          to had a different reaction to
          the idea.

RESUME — Meggs interview.

                    GEOFF MEGGS
          I mean, all the leases that
          were released by Victoria the
          first number of years were
          done overnight. There was no
          pretence at any oversight. And
          people would wake up in the
          morning, and ah, with salmon
                         (MORE)

                    GEOFF MEGGS (CONT'D)
          farms in their view that had
          not been there, ah, a week
          before.

INT. FORUM (MID-1980s) — DAY

Gordon Wilson works for the Sunshine Coast
regional district.

                    GORDON WILSON
          I do not wish to use this as
          a forum for debate on the
          aquaculture industry. We are
          not here to debate the larger
          provincial issues that may be
          at hand.

PHOTO — Mac Richardson stands on the deck of
his ocean-front home. A fish farm is smack in
the middle of his view.

PHOTO — Richardson leads a protest.

                    MAC RICHARDSON (O.S.)
          It's hard for me to be hopping
          mad after fighting with them
          for two years, but more and
          more people are getting mad.
          Commercial fishermen are
          starting to look at the job
          threats to them.

The BC fish farm industry set down its roots on the Sunshine Coast. Early farmers felt that their farms should be as close as possible to Vancouver, which was likely to be the biggest and best market for the fish. (Credit: Juggernaut Pictures Inc.)

Staked Fish Farm Tenures

The government's encouragement for the new industry created a stampede of prospective fish farmers staking farm tenures throughout the Sunshine Coast region. (Credit: Juggernaut Pictures Inc.)

INT. RICHARDSON HOME (MID-1980s) — DAY

He stands in front of the window overlooking the farm.

> MAC RICHARDSON
> Residents are getting mad and fed up with spending thousands of dollars on lawyers to fight with their own government to try and protect their own land.

HEADLINE — The *Fisherman*, "Sunshine Coast puts brakes on aquaculture development," January 17, 1986.

> NARRATOR
> The dust-up between landowners and fish farmers on the Sunshine Coast . . .

HEADLINE — The *Fisherman*, "Sunshine Coast retains salmon farm freeze," January 16, 1987.

> NARRATOR
> . . . pushed the local government to ban aquaculture sites, forcing the farms to more remote locations.

PHOTO — A small, three-pen fish farm is tucked against a remote shore.

PHOTO — A row of zooplankton pods sit in Hidden Basin.

>   NARRATOR
>   This conflict, perhaps, spoke
>   to a larger issue concerning
>   jurisdiction and regulation of
>   the aquaculture industry.

RESUME — McDade interview.

>   GREGORY MCDADE
>   It was clear all along that
>   the ocean was, was something
>   that belonged to the federal
>   government.

MAP — BC south coast.

The area between Vancouver Island and the mainland lights up.

DIAGRAM — The province's jurisdiction of the seabed turns red.

>   GREGORY MCDADE (O.S.)
>   Over time, what took place is
>   there's a territorial problem.
>   A legal theory developed in
>   Canada to say that the province
>   owns the seabed . . .
>   (laughs)
>   . . . between Vancouver Island
>   and the coast of the mainland.

Sunshine Coast residents protest fish farms located too close to residential areas, August 1988. (Photo credit: The Fisherman Publishing Society)

BC's early fish farms were often attached to the shore.
(Photo credit: The Fisherman Publishing Society)

Fish farmers soon discovered that shallow water was easily fouled, was too warm and created problems for the farmed fish because of low oxygen levels. (Photo credit: The Fisherman Publishing Society)

PHOTO — A black-and-white photo of a farm attached to the shoreline.

> **NARRATOR**
> And since the farms were often attached to both the shoreline . . .

INT. OCEAN — DAY

A cable under a farm is attached to the sea floor.

> **NARRATOR**
> . . . and the seabed and required business licences,

FROM ONE FARM TO ANOTHER

> **NARRATOR**
> . . . provincial politicians used this to justify taking the regulatory lead for the industry.

RESUME — McDade interview.

> **GREGORY MCDADE**
> When you look at the, um, the record from, say, 1984 to 1988 . . .

HEADLINE — "Siddon to give provinces control over aquaculture," April 18, 1986.

INT. NEWS CONFERENCE (1988) — DAY

Dave Gillespie and Brian Savage are at a table.

>   GREGORY MCDADE (O.S.)
>   . . . when the federal
>   government was speculating
>   about this, that they were
>   trying to avoid the debate
>   between commercial fishermen
>   and aquaculture operations.

DOCUMENT — "Canada/British Columbia Memorandum of Understanding (MOU) on Aquaculture Development." THIS AGREEMENT made this 6 day of September 1988.

>   NARRATOR
>   Almost two years after the
>   rapid expansion of fish farming
>   on the British Columbia
>   coast, the federal government
>   signed a memorandum of
>   understanding . . .

INT. NEWS CONFERENCE (1988) — DAY

Tom Siddon is signing a document with John Savage. It's the memorandum of understanding that hands over the regulation of aquaculture to the province.

**CANADA/BRITISH COLUMBIA MEMORANDUM OF UNDERSTANDING
ON AQUACULTURE DEVELOPMENT**

THIS AGREEMENT made this   6   day of Sept 1988

BETWEEN THE GOVERNMENT OF CANADA (hereinafter referred to as "Canada" represented by the Minister of Fisheries and Oceans)

OF THE FIRST PART

AND THE GOVERNMENT OF BRITISH COLUMBIA (hereinafter referred to as "British Columbia") represented by the Minister of Agriculture and Fisheries

OF THE SECOND PART

This simple-looking eight-page agreement outlined the cooperation between the Canadian federal government and BC provincial government, and handover of the fish farm industry from the Government of Canada to the Province of BC. Conflicting jurisdictions pushed for such an arrangement since the federal government is responsible for the ocean and the province has jurisdiction over business licensing and the land next to and under the ocean between Vancouver Island and the mainland. This agreement would not solve the jurisdictional impasse, however, and proper oversight of fish farms' impacts on the environment was neglected. (Credit: Government of Canada)

MOU, ANOTHER PAGE — On the words: "5.1 Licensing and Regulation: Provincial" on page 2 of the document.

MOU, LAST PAGE — The signature lines of the document show that Tom Siddon and Brian Savage were the ministers who signed it.

>   NARRATOR
> . . . that officially handed aquaculture over to the provincial government.

FADE OUT:

FADE IN:

INT. HATCHERY (1980s) — DAY

A man scoops some salmon eggs out of a trough and into an aluminum pot.

WIDER — The workers move eggs from one container to another.

ON A TRAY — Now we see the eggs have hatched and are alevins.

INT. HOPE RANCH — DAY

>   BRAD HOPE (O.S.)
> We always ended up, it seemed, with more fish than we could handle and trying to expand and finally . . .

RESUME — Hope interview.

              BRAD HOPE
    . . . a person came to me and
    said, "You know, I have a whole
    oyster operation.

PHOTO — An oyster farmer pulls up a string of oysters to examine them.

            BRAD HOPE (O.S.)
    Why don't we join forces and
    make an aquaculture company?

PHOTO — Brad and two others look into a fry-rearing tub.

            BRAD HOPE (O.S.)
    We can see where this thing
    is going, we can get public
    money, we'll go public." And we
    started Pacific Aqua Foods.

PACIFIC AQUA FOODS BROCHURE — It's promoting the new company.

            BRAD HOPE (O.S.)
    All of a sudden we had close
    to 200 in staff, in marketing,
    running all these different
    operations up and down the
    coast.

PHOTO — Two workers checking the plankton net on a yellow boat.

> BRAD HOPE (O.S.)
> And way before we were ready.

PHOTO — The Port Alice private hatchery.

PHOTO — June and two helpers sort through eggs.

> BRAD HOPE (O.S.)
> And I think what really
> terrified me, one time, was,
> ah, one of the promoters said,

WIDE PHOTO — Top-down view of a rearing tank at the hatchery. Workers are engaged in discussion.

> BRAD HOPE (O.S.)
> . . . he said, "You know what?
> If every fish died, it wouldn't
> matter. We could still sell
> this stock."

RESUME — Hope interview.

> BRAD HOPE
> We all went, that's not what
> it's about. And um, but we
> just, there was no choice. You
> either had, if you wanted to
> keep paying the staff, if you
> wanted to keep, uh, operating,
> you had to grow.

PHOTO — Fish farm staff work in a rudimentary lab.

> BRAD HOPE (O.S.)
> Eventually it became more than we could handle.

PHOTO — Looking down into a brood stock pen.

PHOTO — Incubation trays full of water.

> BRAD HOPE (O.S.)
> So we had a major accident. Hatchery water quit. Virtually the hatchery year class was wiped out. Eggs and everything.

PHOTO — All the ponds are empty.

PHOTO — A view of the support building from the wharf.

> BRAD HOPE (O.S.)
> It was just a company takeover. And we were bought out and, and ah, it moved on.

RESUME — Hope interview.

> BRAD HOPE
> Bought out for, of course, pennies on the dollar. But, I, I, you know, I realize it was all, ah, much of it was orchestrated.

PHOTO — The Hidden Basin pens at dusk.

ANOTHER PHOTO — The pens look empty, quiet.

>   NARRATOR
> The takeover of Tidal Rush and
> Pacific Aqua Foods was not a
> special case by any means.

HEADLINE — The *Fisherman*, "Food multinationals are taking over B.C. salmon farms," March 16, 1990.

>   NARRATOR
> Almost every small fish farm
> company on the west coast
> was taken over just as a DFO
> manager predicted.

HEADLINE — The *Fisherman*, "Big companies rule aquaculture in B.C.," September 21, 1992.

>   NARRATOR
> He stated in the early 1980s
> that . . .

THE WORDS — "this industry will be controlled by large foreign companies." lift off the page.

>   NARRATOR
> . . . "This industry will be
> controlled by large foreign
> companies." It didn't take
> long.

EXT. WEST COAST FISH FARM (AERIAL) — DAY

Establish. A farm with round pens.

EXT. WEST COAST FISH FARM (AERIAL) — DAY

Establish. Another fish farm with a single row of square pens.

> NARRATOR
> In less than a decade, the west coast industry shrank from over 100 companies to 20.

HEADLINE — The *Fisherman*, "Ireland freezes foreign fish farms," July 18, 1986.

> NARRATOR
> And this consolidation was happening overseas as well.

INT. VANCOUVER APARTMENT — DAY

Establish. Interview. Bartlett Naylor is an economist for the wild salmon trust.

TITLE: Bartlett Naylor, Consultant, Corporate Governance.

> BARTLETT NAYLOR
> When I began the research, there were roughly five or six companies that controlled 80 or so percent of world production
> (MORE)

                    BARTLETT NAYLOR (CONT'D)
               in every geography. Norway,
               Chile, Canada, UK.

MAP — BC south coast.

Coloured dots represent each fish farm
company's tenures. They morph from many
colours and many companies to predominantly
three colours, representing three companies.

                    BARTLETT NAYLOR
               And even as I was ah
               researching this, ur, or in
               the subsequent couple of years,
               that collapsed to really two or
               three companies.

MONTAGE — Several salmon farms.

                    NARRATOR
               Even though most fish farm
               tenures are now held by just
               three companies, concerns
               about the environmental impacts
               increased.

EXT. LARGE CORPORATE FARM — DAY

Atlantic salmon jump in the pen.

                    CRAIG ORR (O.S.)
               We brought some of the
               researchers to this coast . . .

INT. CRAIG ORR'S OFFICE — DAY

Establish. Interview.

> CRAIG ORR
> . . . like Paddy and took them out in the Broughton to look around. These are the European researchers. They were staggered by the size of our salmon farm.

EXT. LARGE FISH FARM — DAY

The size of the farm is impressive.

> CRAIG ORR (O.S.)
> Much, much larger. Not quite as numerous, but much larger than, than the farms that you see in the European Union countries.

EXT. LARGE FISH FARM (AERIAL) — DAY

Also impressive in size.

> JOHN VOLPE (O.S.)
> Would there be a problem with one salmon farm on this coast? Two? No. Three, no. Twenty, maybe. Fifty,

INT. VOLPE'S OFFICE — DAY

Establish. Interview. John Volpe is a professor at the University of Victoria, BC.

TITLE: Dr. John Volpe, Principal Researcher, SEG Group, University of Victoria.

              JOHN VOLPE
  . . . perhaps. A hundred and twenty, like we got now, clearly there is. And so, where is, where is, the, you know, where is the trade-off? Where is that breaking point where, you know, economic benefit begins to trump ecological and social, for that matter, benefit?

INT. OCEAN — DAY

Herring swim through an otherwise vacant ocean.

              NARRATOR
  The question is, do we have enough information to know how many fish farms could have a significant impact on wild fish?

EXT. NET PEN — DAY

Herring spawn on the netting of a farm net pen.

                    NARRATOR
          There is evidence that the
          number might be much smaller
          than previously thought.

MAP — BC coast.

The water areas are marked in traditional blue. Then the southernmost part of the coast near the United States border turns red.

                    NARRATOR
          In 1978, the Department of
          Fisheries and Oceans reported
          sharp declines in herring
          populations in their southwest
          management areas.

RESUME MAP

The Salish Sea now lights up in red. Then, the areas surrounding the northern half of Vancouver Island light up in yellow. And the areas further north are still green.

                    NARRATOR
          This was followed by sharp
          decreases in the Salish Sea
          the next year, with moderate
          decreases further north.

The red area begins to expand north, and the yellow zones slowly encroach on the green.

>   NARRATOR
>   And just three years later, by
>   1982, the areas of moderate
>   decline were now sharply
>   declining. And areas even
>   further north that were
>   previously unaffected were now
>   experiencing moderate declines.

                                        FADE TO:

HEADLINE MONTAGE — The *Fisherman*, "Herring landings the largest since sixties," January 28, 1977; "UFAWU voices fishermen's concern: Herring fishery mismanaged," April 1978; "West coast herring 'in trouble,'" April 22, 1982; "Herring stocks at '64 level on entire south, central coast," October 18, 1985.

>   NARRATOR
>   Herring stocks on the west
>   coast went from historic highs
>   in 1977 to mismanaged in 1978,
>   to being in trouble by 1982,
>   to the record lows of the mid-
>   1960s. A population crash that
>   took just six years.

                                        FADE TO:

OVER BLACK

RUSSELL MEMORANDUM — Quickly moving in on the word "Confidential." Then panning down to the date: August 11, 1988.

> NARRATOR
> A once confidential DFO memo written three years later in 1988 . . .

FLIPPING THE PAGE: The words "in 1985 there were . . . fewer than 5 net pen sites" lift off the page.

> NARRATOR
> . . . observed that "in 1985 there were fewer than five net pen sites."

MAP — BC's Sunshine Coast area.

Dots identify the location of the few farms that existed at that time.

> NARRATOR
> And all of the BC farms were located along the Sunshine Coast.

RESUME — The Russell memorandum.

THE WORDS — "a traditionally spawned section of coastline between two salmon net pen facilities has not been spawned since the farms became operational" lift off the page.

                    NARRATOR
    The author also reported a rather disturbing fact that "a traditionally spawned section of coastline between two salmon net pen facilities has not been spawned since the farms became operational."

MONTAGE — Flyovers of one fish farm after another.

                JOHN VOLPE (O.S.)
    You don't need to go out and actually collect data to make the prediction that those elements of the ecosystem proximate to the farms are going to be the ones that feel the effects first and potentially to the greatest magnitude. As the mushroom cloud expands, you know, it will encompass greater and greater amounts of area. And I always find it interesting when . . .

RESUME — Volpe interview.

                    JOHN VOLPE
          . . . critics stand up and, and,
          um, try to explain away these
          kinds of things like it's,
          well, like it's not high school
          science.

                                    FADE TO:

MAP — Animated Herring Spawning Activity
British Columbia south coast.

As the years from 1980 to 2012 roll over on
screen, the changing spawning pattern for
herring is mesmerizing and disturbing. The
spawning areas are shrinking rapidly.

                    NARRATOR
          Annual surveys by the
          Department of Fisheries
          and Oceans show that the
          disappearance of herring
          between the two farms was not
          an isolated event. Since 1980,
          herring spawning had declined
          dramatically over the entire
          south coast.

EXT. RIVER — DAY

Establish a wild river.

INT. RIVER — DAY

Eulachon swim up the river.

> NARRATOR
> Another small pelagic species, the eulachon, lives in the ocean and spawns in west coast rivers. Suddenly, they began to disappear in the late 1980s, and by 1994 the Fraser River run had declined by 90 percent.

EULACHON GRAPH — The eulachon populations of the Fraser decline over the 1980s and 1990s.

BC RIVER MAP — BC coast, showing major rivers.

One by one, we see the rivers that once supported large eulachon runs from south to north: Fraser, Kingcome, Owikeeno, Bella Coola, Atnarko, Kimquits, Kitilope, Kitimat, Skeena, Unuk and Stikine. The red lines indicate they are populations in trouble.

> NARRATOR
> And once large populations on at least ten other rivers had runs so small by the year 2000, scientists fear some may now be extinct.

HEADLINE — The *Georgia Strait*, "Mystery on the River," December 3, 1993.

Food multinationals are taking over B.C. salmon farms

After just over a decade, consolidation of the fish farm industry was underway. The *Fisherman* reported in this article from March 16, 1990, that large multinational food companies were buying up farms that were struggling with a new disease, marine anemia. (Credit: The Fisherman Publishing Society)

Fish farms continue to face opposition and salmon kills

Algae blooms, storm damage, fish escapes and mounting protests against Atlantic salmon rearing in lakes continue to plague a shaky B.C. aquaculture industry.

On the West Coast of Vancouver Island in early September an estimated 250,000 farmed salmon were killed by a sudden algae bloom. The microscopic algae are poisonous to fish when they reproduce rapidly and drift into net pens.

Western Harvest Seafarms in Zeballos lost about 100,000 chinooks and Intercan Resources III Ltd. with 60 pens in Kyuquot Sound also had massive fish kills. The dead fish were dumped offshore west of Brooks Peninsula. There is no indication what the effect of dumping tons of dead salmon at sea will be, though the government approved the dumping.

Many fish farms re-established on the West Coast of Vancouver Island after experiencing algae problems in the more protected waters of Georgia Strait. It was thought the waters on the West Coast were less susceptible to algae blooms.

Escapes from farms are a problem on the West Coast too. In December 1990 Intercan III lost 350,000 chinooks during a winter storm near Kyuquot. The farm had gone into its location without DFO approval. Local fishermen were concerned that the escaped fish would breed with native fish, disrupting the genetic base of the wild stocks.

Atlantic salmon farming is rapidly increasing in British Columbia, even though the effects of Atlantic salmon rearing and intermingling with wild Pacific salmon are largely unknown.

The UFAWU believes that the government and the aquaculture industry should have proven that there would be no harm to native stocks or the environment before allowing any kind of Atlantic salmon production to go ahead.

In Norway, infestation of Atlantic salmon by the gyrodactylus salaris parasite has forced the government to consider sterilizing (killing all fish) in 70 streams in an attempt to eradicate it. The fish disease furunculosis reached epidemic proportions in Norway's aquaculture industry and rapidly spread to wild streams.

Farmed Atlantic salmon in Norway have escaped and intermingled with wild salmon in streams, threatening the genetic diversity of wild salmon. Norway has now set up "gene banks" to protect the remaining genetic diversity of their wild stocks.

John Willow, aquaculture development officer at the ministry of Agriculture and Fisheries says that Atlantics are preferred by fish farmers. "They grow bigger faster," he said. "It's basically a premium cow. It's cheaper to produce a pound of Atlantic than a pound of Chinook."

Willow says that there are 101 operating producing salmon farms on the West Coast now, controlled by about 35 companies.

About 20,000 metric tonnes of farmed salmon is now produced annually in B.C.

Despite concern by Ministry of Environment officials, pen rearing of Atlantic Salmon smolts is already in place at Lois Lake near Powell River and Georgie Lake near Port Hardy. A push by the aquaculture industry to allow fresh water rearing of Atlantics near Tofino was defeated earlier this year by a coalition of sports and commercial fishing industry groups. They were concerned about the spread of disease from the farmed fish to native coho and sockeye in the lake system. Two other B.C. lakes have been identified as possible Atlantic rearing sites.

Commercial fishermen have been alarmed at an increased number of Atlantic salmon they have caught this year.

Ron Ginetz, regional aquaculture co-ordinator for DFO says his department has reports of increased landings of Atlantics this year. "Each plant might get five or six after a fishery," he said.

Salmon farms are currently producing about 80 percent chinook and coho salmon and only 20 percent Atlantics. Ginetz predicts a shift to where B.C. farmed production will be about one-half chinook and one-half Atlantics. "Now that we have a greater interest in Atlantic salmon we have to make sure there isn't a cause for concern," he said. "We don't believe Atlantic salmon are a significant problem to Pacific salmon and we're trying to gather additional information to confirm that."

Fishers who catch an Atlantic salmon should call Ginetz at 666-3152.

Big companies rule aquaculture in B.C.

Seven companies control 80 percent of farmed salmon production in B.C., much of it Atlantic salmon.

The three biggest producers are B.C. Packers, Pacific Aquafoods, (a branch of National Sea Products), and Stolt Seafarms, a branch of Stolt Tankers and Terminals, a huge Norwegian multi-national corporation.

John Willow of the ministry of Agriculture and Fisheries says that from a high of 90 companies when salmon farming started in B.C. in the early 1980s, there are now between 30 and 35 companies operating in the province.

In 1992, the *Fisherman* reported more trouble with disease and fish farm mortalities. There were 101 operating farms, but just 7 companies controlled 80% of the Atlantic salmon production. (Credit: The Fisherman Publishing Company)

                    NARRATOR
          Even the majestic white
          sturgeon began to die in the
          early 1990s in large numbers,

PHOTO — A large sturgeon lies dead in the muck.

                    NARRATOR
          . . . crushing the population
          and rendering it a species at
          risk.

PHOTO — Another sturgeon lies dead on the edge of the Fraser River.

INT. RIVER — DAY

A giant white sturgeon swims slowly through the frame.

PHOTO — A dead sturgeon, with a red belly, lies in shallow water.

                                        FADE TO:

EXT. RIVER — DAY

Establish.

A STEELHEAD — Tight on the fish's head. It's alive.

HEADLINE — The *Fisherman*, "Steelhead in crisis: Who is to blame?" April 22, 1991.

INT. CRAIG ORR'S OFFICE — DAY

>> CRAIG ORR (O.S.)
> You know, the whole east coast of Vancouver Island steelhead population has really crashed. And, ah,

RESUME — Orr interview.

>> CRAIG ORR
> . . . you know, there is certainly a very strong possibility that part of that could be related to impacts on juvenile steelhead migrating past salmon farms.

HEADLINE — The *Fisherman*, "What's happening to the coho?" March 23, 1998.

>> NARRATOR
> And the story is the same for Pacific salmon species. Their populations, just like herring,

MAP — Animated zones, BC coast.

It's the same map of the BC coast that showed herring declines. And a similar thing happens to salmon. Red zones begin in the south and move north.

                    NARRATOR
          . . . began declining in the
          south and moved north. Coho
          and chinook started to decline
          first in the early 1980s,
          with pink, chum and sockeye
          declining just a few years
          later.

HEADLINE — The *Fisherman*, "Low returns create poor season in 1992," August 24, 1992.

                    MARTY KRKOSEK (O.S.)
          If you look at where the salmon
          farms are located,

INT. ECHO BAY LAB — MORNING

Establish. Interview. At the time, Dr. Krkosek was a PhD candidate doing research on the impact of fish farms on wild Pacific salmon.

TITLE: Dr. Martin Krkosek, Assistant Professor, Dept. of Ecology, University of Toronto.

                    MARTY KRKOSEK
          . . . they're right on most
          of the wild salmon migration
          routes, whether it's in the
          Broughton Archipelago or in
          the waters off of Campbell
          River . . .

MAP — BC's Inside Passage.

The salmon migration routes through the Inside Passage pass through the channels where fish farms are located.

> MARTY KRKOSEK (O.S.)
> . . . where all the Fraser stocks come through. Ah . . . They're situated in such a way that cause a collision between the wild and the farm salmon.

RESUME — Morton interview.

> ALEXANDRA MORTON
> If I then extrapolate to the BC coast, I can only imagine that wherever herring and salmon farms collide, it's very likely the same dynamic is going on. And so DFO argues with me that this might be an inconsequential, small problem. But that's not the way I see it as a biologist who's been looking at this one place for 20 years.

INT. OCEAN — DAY

Herring are spawning on shoreline vegetation.

                    NARRATOR
          Herring, for example, migrate
          inshore to spawn in the
          intertidal zones and then
          return offshore to their
          feeding grounds every year.

ANIMATED MAP — Herring migration on the BC
coast.

Pulses of herring schools go inshore to spawn
and then return to their feeding grounds
offshore.

                    NARRATOR
          And because they don't die
          after spawning, each individual
          herring has the potential to
          pass by fish farms at least
          twice a year for every year
          they are alive.

INT. OCEAN — DAY

Migrating sockeye smolts make their way along
a shoreline.

                    NARRATOR
          While salmon, on the other
          hand, might only pass by the
          farms twice.

EXT. FISH FARM — DAY

A very large farm sits in a remote channel.

                    NARRATOR
          Once when they migrate out to
          sea,

EXT. RIVER — DAY

Sockeye salmon mill around in a spawning
channel.

                    NARRATOR
          . . . and a second time when
          they return to spawn.

INT. OCEAN — DAY

Mature salmon fight the current in the
estuary, silver and fresh from the ocean.

                    NARRATOR
          And it's not just herring and
          salmon that have migratory
          routes that can pass by the
          farms. Where they go, how often
          they go and what they eat . . .

INT. RIVER — MOMENTS LATER

The sockeye salmon chase each other for
space.

                    NARRATOR
          . . . will change the risk
          profile of a species or stock
          of fish. And these life cycle
                         (MORE)

> NARRATOR (CONT'D)
> differences will change the
> speed at which a particular
> population might decline from
> fish farm contact.

HEADLINE — The *Fisherman*, "Huge Skeena run spawns uproar," October 18, 1985.

> NARRATOR
> Fish farmers, however, are
> quick to point out that some
> runs have done exceptionally
> well.

HEADLINE — The *Fisherman*, "Horsefly sockeye run a 'miracle,'" August 16, 1985.

> NARRATOR
> These very large runs baffled
> DFO scientists . . .

HEADLINE — The *Fisherman*, "Mixed signals for '85: Processors reap sockeye bonanza," May 17, 1985.

> NARRATOR
> . . . since the recruitment,
> the number of adult fish
> returning from a pair of
> spawners, was much, much higher
> than normal.

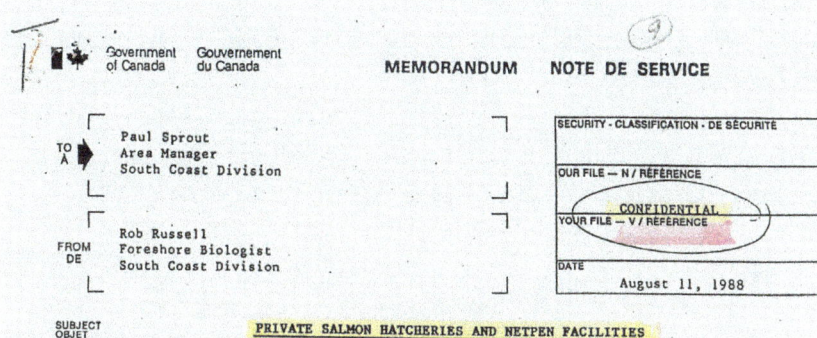

Government of Canada	Gouvernement du Canada	**MEMORANDUM**	**NOTE DE SERVICE**

TO / A: Paul Sprout, Area Manager, South Coast Division	SECURITY - CLASSIFICATION - DE SÉCURITÉ
	OUR FILE — N / RÉFÉRENCE
	CONFIDENTIAL
FROM / DE: Rob Russell, Foreshore Biologist, South Coast Division	YOUR FILE — V / RÉFÉRENCE
	DATE: August 11, 1988

SUBJECT / OBJET: PRIVATE SALMON HATCHERIES AND NETPEN FACILITIES — SOME SERIOUS CONCERNS

In light of recent allegations by the U.F.A.W.U. and televised responses from the Aquaculture Industry, we would like to bring the following to your attention.

Since 1984 the number of private salmon hatcheries and netpen sites have increased dramatically in B.C. and particularly in the South Coast Division. In 1984 there were no private salmon hatcheries and fewer than 5 netpen sites. As of June 1988, 47 hatcheries have been built or approved for South Coast Division and 178 netpen sites have been approved for B.C. (170 of the sites are in S.C.D.).

As you are aware, every hatchery and netpen facility is vetted through a series of referrals involving three levels of Government and the concerned public. With the occasional exception, the referral system does an adequate job of addressing D.F.O.'s concerns relating to wild fish and their habitats. It is the "after approval" impacts of private salmon husbandry, not addressed by the referral system, which are becoming a serious concern for many S.C.D. staff.

What follows is an abbreviated list of documented problems arising from private salmon facilities in South Coast. Many of these have been brought to our attention by employees of the operations themselves or by private citizens. In most cases, a more detailed account of each concern is available from an investigating Fishery Officer or habitat biologist.

a) JUVENILE SALMON

1. Salmon fry and smolts have been observed swimming through several netpen facilities, primarily in the spring. Interviews with on-site staff have confirmed that salmon fry are eaten by farmed salmon. Interviews with processing plant staff have also confirmed the presence of salmon fry in the stomachs of farmed fish.

2. Juvenile salmon gill themselves on the predator net at one farm site. A diver at the farm confirms the removal of as many as four fish a day killed in this manner.

.../2

HEADLINE — The *Fisherman*, "DFO baffled by Barkley sockeye, chinook returns," August 21, 1987.

HEADLINE — The *Fisherman*, "Surprise pink return in north baffles DFO," August 10, 1983.

                   NARRATOR
    But are these anomalies
    evidence that the fish farms
    are having no effect?

GRAPH — Henderson Lake Sockeye abundance by year. Abundance spikes then crashes.

                   NARRATOR
    If you look at the profile of
    each stock,

GRAPH — Skeena River Sockeye abundance by year. Abundance spikes then crashes.

                   NARRATOR
    . . . these sudden spikes in
    abundance were usually followed
    by poor recruitment and a steep
    population decline.

GRAPH — Long Lake sockeye abundance by year. Abundance spikes and then crashes.

                    NARRATOR
          And many populations have not
          recovered. So what does this
          mean?

ANIMATED MAP — Two competing BC salmon
populations.

Migration routes of two salmon populations
overlap in a feeding ground. One population
is suddenly smaller and the other one gets
larger.

                    NARRATOR
          In simple ecological terms,
          it means the ecosystem was
          suddenly out of balance. An
          increase in one population
          was probably the result of a
          decline in another population.
          The sudden lack of competition
          for food or disappearance of a
          predator resulted in increased
          survival for the benefiting
          population.

INT. OCEAN — DAY

A school of adult fish mill around near the
mouth of a river.

Huge Skeena run spawns uproar

The loss of at least one million Skeena River sockeye which passed the commercial fishery but were unable to spawn in the Babine Lake spawning channels is generating a political storm in northern B.C.

At least one of the spawners made it all the way to the House of Commons desk of Prime Minister Brian Mulroney, who was absent Oct. 16 at the Commonwealth Conference in the Bahamas.

Skeena NDP MP Jim Fulton provoked an uproar in the House when he produced the dripping fish and threw it on the prime minister's desk in the course of a question to acting fisheries minister Eric Nielsen.

Nielsen had earlier pledged every effort would be made to save the surplus spawners, although it was not clear what he had in mind. By Oct. 15, observers at the Fulton and Pinkut channels reported the lake was red with sockeye packed in a solid mass at the gates of the stream.

Indian organizations representing commercial fishermen demanded the resignation of fisheries northern director Eric Kremer over the incident while inland tribal councils pointed to the overspawn as proof of the viability of an inland commercial fishery.

For his part, Kremer terms the loss the result of a "classic mixed-stock fishery" in which strong enhanced runs are mingled with smaller or threatened runs of chinook, coho and steelhead.

The size of the disaster became obvious in late September when fisheries officers at Babine Lake counted more than two million spawners past the counting fence. Pre-season escapement targets were only 1.3 million, meaning as many as one million fish would spawn in the lake.

The department had expected a run of only 2.7 million, Kremer said Oct. 5, and developed a system of four-day fisheries that would allow sustained harvest of the two spawning channel runs while protecting some 48 other unenhanced systems.

In fact, the total run hit a 100-year peak of 4.7 million, of which 2.5 million were harvested. The extra escapement boosted the surplus in the lake from a forecast 300,000 to at least one million.

The Northern Native Fishing Corp. and the Nishga Tribal Council demanded Kremer's resignation for his failure to allow the commercial harvest of the surplus.

Gitksan-Wet'suwet'en Tribal Council president Neil Sterritt, however, said the fish should have been used for his organization's proposed inland commercial fishery.

Kremer said Oct. 4 that there wasn't enough time to install the facilities necessary for a harvest at Babine Lake even if a market could have been found. As it was, all streams on the Babine had optimum escapement.

Skeena NDP MP Jim Fulton, who called in the House of Commons for some action to use the fish, said later the department should have allowed more commercial fishing and needs to develop ways to do "more accurate stock differentiation at the mouth of the Skeena."

"If we don't start going that road, it's clear some measures must be taken for in-river harvest."

UFAWU northern representative Jim Rushton said the union believes some changes are necessary in fishing patterns to allow full harvest of the enhanced runs without damaging weaker stocks.

THE FISHERMAN — OCTOBER 18, 1985/15

Many salmon runs were experiencing spectacular declines in recruitment by the mid-1980s. This means there were far fewer than normal adult spawners per pair of spawners. With sockeye, the historical norm was six adult returning fish from each pair of spawners. But during this time, some runs would experience much higher numbers of spawners, which baffled Fisheries scientists. This phenomenon indicated that the population with higher recruitment was experiencing a temporary lack of competition for food in the environment. The spike was soon followed by a crash when the population became the next one to be infected with newly introduced diseases like viral hemorrhagic septicemia (VHS), a virus that causes internal bleeding. (Credit: The Fisherman Publishing Society)

Spectacular runs were not limited to sockeye, but all species of salmon would have exceptional years of high recruitment followed by just as spectacular declines. This meant that the disease affecting wild fish populations was not species-specific and could infect many different species of fish. VHS is just one virus that is lethal to many different species. (Credit: The Fisherman Publishing Society)

                    NARRATOR
          But that sudden benefit meant
          that whatever took down the
          competitor was nearby and was a
          sign of bad things yet to come.

                                        FADE TO:

EXT. FISH FARM — DAY

A boat enters frame. On board is a person dressed as a salmon and holding a sign that says "Just say no to farmed salmon."

SUPERSCRIPT: Get Out Migration Protest — Day 4

A YOUNG WOMAN — She's wearing a wet suit. She somersaults off a boat and into the water.

WIDER — She swims along the perimeter of the fish farm.

ON ANOTHER BOAT — A few moments later.

MORE PROTESTERS — They're standing behind a sign that says "Marine Harmfest."

EXT. CHANNEL — MOMENTS LATER

Protesters in several boats wave flags.

EXT. BOAT — DAY

Alexandra Morton smiles as she looks out at the other protesters.

EXT. QUADRA ISLAND DOCK — DAY

Alexandra Morton arrives at the island. People from several First Nations and the media are there.

A NORWEGIAN FLAG — The words on it read "Norway Open-net Fish Farms Out."

ANOTHER SIGN — The words on it read "I don't want to be a dead mort, I want to be a living wild salmon!!"

EXT. QUADRA ISLAND PARK — DAY

Alexandra arrives to applause from the crowd.

FADE TO:

EXT. STREAM — DAY

Clean water flows over a bed of gravel.

ANOTHER SHOT — Light flickers through the water, dancing on the rocks and pebbles that lie on the bottom.

As part of the Get Out Migration, the protesters left their walk down Vancouver Island to protest for one day at a fish farm near Quadra Island. (Photo credit: John Preston)

Quadra Island residents meet with protesters at the island's public dock. Alexandra Morton and other members of the community form a circle to listen to a speaker. (Photo credit: John Preston)

CLOSE UP ON SALMON EGGS — The eggs are lying in the gravel. A developing fry spins in the egg.

CLOSER — An eye or two stare back at us.

                    NARRATOR
          One of the biggest challenges
          facing the new industry was
          acquiring brood stock. The
          farmers needed eggs to grow
          fish.

AN EGG — It bursts open and a salmon fry hatches out.

                    BRAD HOPE (O.S.)
          Where could we get eggs? It
          just got more and more exciting,
          and of course . . .

RESUME — Brad Hope interview.

                    BRAD HOPE
          . . . there was no rules or
          regulations with the province.
          It was just, they had never
          really run into it except in
          a very few instances. So they
          said, yeah, if there's excess
          eggs in the hatchery we could
          sell them.

INT. HATCHERY — DAY

A worker fills an aluminum pot with salmon eggs. He scrapes off the excess.

          BRAD HOPE (O.S.)
    So, ah, well, we said ok, let's
    start going and that's, that's
    where it took off.

TIGHTER — The pot is full. It's lifted out of the water and water drains out of the holes in the bottom.

          BRAD HOPE (O.S.)
    So, we went in actually to get,
    I think about 20,000. And then
    thought,

ON THE POT — The worker dumps the eggs into a plastic tub.

          BRAD HOPE (O.S.)
    . . . well, they're so cheap,
    why don't we get 40,000, and
    then when we were there,
    somebody said well, we've
    got lots, why don't you take
    80,000?

HATCHERY WORKERS — continue to sort eggs into tubs.

HEADLINE — The *Fisherman*, "Fish farms: a bad idea getting worse," August 16, 1985.

                    NARRATOR
          After a humble beginning where
          the early farmers were taking
          less than 100,000 eggs, by the
          middle of the decade . . .

PAN DOWN — The words "More than one million
chinook eggs will find their way out of
public hatcheries in 1985 to private farms"
lift off the page and are highlighted.

                    NARRATOR
          . . . the Department of
          Fisheries and Oceans promised
          to supply eggs from the
          taxpayer-funded Salmon
          Enhancement Program.

PHOTO — About two dozen eyed eggs are on a
counting table.

ANOTHER PHOTO — The frame is full of eggs,
but some of them have hatched and are now
alevins.

                    NARRATOR
          Millions of eggs that were once
          earmarked for the enhancement
          of wild salmon were being sent
          to farmers as seed stock . . .

EXT. EARLY FISH FARM — DAY

From the entrance to a support shed onshore,
the farm stretches out into a gloomy ocean.

A LARGE CREW — of workers are busy preparing the nets.

>   NARRATOR
>   After securing the egg supply,
>   the next challenge faced by the
>   farmers was to turn these eggs
>   into mature fish.

Someone pulls fish out of a pen with a long-handled dip net. They are already dead. The worker drops the dead salmon on the wharf. Something is wrong.

TIGHT ON DEAD SALMON — A worker is using a dip net to pull dead salmon out of a pen and dumping them on the dock.

EXT. WHARF — DAY

A tote is full of dead salmon. The workers toss the dead fish into another tote.

SUPERSCRIPT: Farm salmon die-off, mid-1980s.

>   BRAD HOPE (O.S.)
>   The difficulties that we
>   experienced when we first
>   started. Ah, when we first put
>   them in salt water, how long
>   do we have in this interview,
>   because it seems really obvious
>   now that . . .

The workers push the tote toward shore.

IN THE PEN — There are more dead fish floating around and lots dead at the bottom of the net.

>					BRAD HOPE (O.S.)
>			. . . you would know about
>			otters and herons and seals.

ANOTHER FARM — Lots of nets cover the pens.

>					BRAD HOPE (O.S.)
>			And there was no coating that
>			you could put on them that
>			would keep mussels off, so you
>			would put a net and ah, within
>			a month it weighed tons.

TWO WORKERS — They're working in a net pen in the setting sun.

>					BRAD HOPE (O.S.)
>			Plankton blooms, like there was
>			Chaetoceros, a spiky little
>			thing . . .

IN A TOTE — A frozen salmon is lying with its mouth open, as if it had been screaming when it died.

>					BRAD HOPE (O.S.)
>			. . . that would get in the
>			gills and kill them.

RESUME — Meggs interview.

A giant white sturgeon was found floating in the ocean near the mouth of the Fraser River (photographed October 19, 1993). The fish appeared to have red, sunburnt skin, which is a typical symptom of VHS. Sturgeon often feed on dead salmon in the river, and this one likely contracted the disease by consuming dead wild salmon infected with VHS. (Photo credit: Marvin Rosenau)

Marvin Rosenau examines a dead Fraser River sturgeon on August 29, 1994. This sturgeon was one of many of the giant fish that died over several years. Most had the red skin coloration typical of VHS. (Photo credit: Marvin Rosenau)

                    GEOFF MEGGS
          I mean, there was, there was
          a significant problem during
          the '80s of illegal disposal
          of mass die-offs of fish. Ah,
          they were dropped in ravines,
          they were taken to landfills.
          There's a hilarious but
          disgusting story by Howie White
          who used to run the Pender
          Harbour dump of about . . .

EXT. DUMP SITE (1980s) — DAY

A row of large totes sit at the edge.

                    GEOFF MEGGS (O.S.)
          . . . having to dig a massive
          trench to bury, you know,
          hundreds of tons of salmon.

EXT. DUMP SITE (1980s) — DAY

The crane operator pulls a lever and tips the totes, dumping out a disgusting slop of dead fish.

                    IAN GILL (V.O.)
          This is fish farming most foul.

TIGHT ON THE CARCASSES — They look like a brown semisolid slop.

                    IAN GILL (V.O.)
               These are farm-raised fish
               that didn't make it. Instead of
               ending up on a plate somewhere,

A WORKER — hooks up a loop for the crane to
tip more totes of rotting fish.

                    IAN GILL (V.O.)
               . . . they're being dumped in
               a landfill site. Thousands
               of them will end up here this
               season.

TIGHT ON A TOTE — The dead fish are a brown
pile of muck.

                    IAN GILL (V.O.)
               A vile, putrefying slop, that
               should you be able to smell it,
               would put you off eating for
               life.

THE NEXT TOTE — It's tipped over and the
rotting fish contents spill into the deep
trench.

                    IAN GILL (V.O.)
               It's disgusting. So much so,
               that even the seagulls won't
               touch it.

IN THE TRENCH — It's full of dead and
decaying salmon.

Farm fish die-offs are common at fish farms. The fish cannot be sold, and during the early days rotting carcasses were dumped into large trenches. Some farms later got permission to dump their dead fish at sea. (Photo credit: The Fisherman Publishing Society)

DISSOLVE TO:

PHOTO — June Hope and a helper are collecting milt from a chinook.

> NARRATOR
> The difficulties fish farmers were having with raising Pacific salmon . . .

PHOTO — A farmhand holds two mature chinook by the tail.

> NARRATOR
> . . . pushed them to look for stocks that might be more disease-resistant and would grow faster.

HEADLINE — "Eggs from 9 Chinook stocks taken for farmers' gene pool," November 20, 1987.

> BRAD HOPE (O.S.)
> They wanted us to take eggs from a number of different rivers . . .

RESUME — Brad Hope interview.

> BRAD HOPE
> . . . and hatch them and grow the fish and see what, kinda see what difference there was, and they wanted to know is if a
> (MORE)

                    BRAD HOPE (CONT'D)
               strain got fished out, could we
               supply eggs and spawners . . .

EXT. TIDAL RUSH FARM (1980s) — DAY

June and a helper pour eggs out of a washtub and into a pitcher.

                    BRAD HOPE (O.S.)
               . . . in the future to try and
               rebuild it. And so, we set up
               separate net pens to hatch
               out the eggs, and reared them
               separately.

EXT. FRY-REARING POND — DAY

A pair of hands lifts a small dip net and dumps a couple of smolts into a tall glass jar.

PULLING BACK — A young woman assesses her sample.

                    BRAD HOPE (O.S.)
               And what was really interesting
               was how different they were,
               how differently they grew.

SMOLTS — They're wriggling in the glass jar.

> BRAD HOPE
> How quickly they grew, how they adapted to, to different fish feeds and . . .

EXT. NET PENS — DAY

A worker scoops some feed out of a garbage can filled with fish feed pellets.

> BRAD HOPE (O.S.)
> . . . one of the things I remember is going in the pens,

He tosses the food into the pen and the fish thrash at the surface.

> BRAD HOPE (O.S.)
> . . . and some of them, you would step onto the dock and as soon as they sensed there was somebody there, they would be up looking for feed.

LOOKING DOWN — The fish continue to feed on the pellets.

> BRAD HOPE (O.S.)
> Others, you would step on as soon as you're on. You never saw a fish. They were down at the bottom.

RESUME — Hope interview.

                    BRAD HOPE
          But in, again, in small net
          pens that were really closely
          attached we saw very different
          levels of die-off and survival.

INT. CHINOOK NET PEN (EARLY 1980s) — DAY

Chinook salmon mill around in the pens.

                    NARRATOR
          But experimentation was not
          just limited to growing out
          different stocks. Farmers also
          experimented with mixing stocks
          together to try and create a
          better farm fish.

                    BRAD HOPE (O.S.)
          We saw that as a . . .

RESUME — Hope interview.

                    BRAD HOPE
          . . . real research project for
          the future is to be able to get
          all these different strains and
          raise them right through, and
          ah, but we never did because
          there was so many problems
          with, uh, chinook.

RESUME CHINOOK PEN

                    NARRATOR
          Chinook salmon mature at
          different rates, and in the
          wild come back to spawn after
          two, three, four, five years.
          This uneven maturation rate
          created another problem for
          farmers looking to deliver a
          uniform product.

                                    FADE TO:

EXT. TIDAL RUSH FARM — DAY

The crew are trying to separate brood stock
in a pond on land.

                    BRAD HOPE (O.S.)
          We tried for years and years to
          bring Atlantics in, and it was
          very frustrating because there
          was no Atlantics allowed on the
          west coast of Canada . . .

INT. GOVERNMENT HATCHERY — DAY

A worker pours eggs from a saucepan into a
covered channel.

                    BRAD HOPE (O.S.)
          . . . at all. So we said, look
          we'll do everything, we'll do
          all kinds of isolation,

TIGHT ON TUBE — The hatchery worker shakes the eggs into a rearing channel.

> BRAD HOPE (O.S.)
> . . . we'll do all the, the disease control.

> OTTO LANGER (O.S.)
> Then bringing Atlantic salmon . . .

RESUME — Langer interview.

> OTTO LANGER
> . . . some of the directors of Fisheries in Victoria and the director general of DFO said, well, that's an exotic species.

HEADLINE — The *Fisherman*, "Current policies to import Atlantic salmon eggs amount to an act of biological insanity," April 18, 1986.

> OTTO LANGER (O.S.)
> We don't want it here. The Ottawa direction was, take it . . .

INT. HATCHERY — DAY

A worker takes a tray of newly hatched salmon over to a growing tub.

                    OTTO LANGER (O.S.)
          . . . and just make certain
          they're allowed to farm, and
          it's your job to make certain
          it's done safely. Well, how do
          you safely bring in Atlantic
          salmon . . .

THE WORKER — pours newly hatched eggs through
a screen filter.

                    OTTO LANGER (O.S.)
          . . . and keep out exotic
          diseases that can destroy our
          fishery on this coast?

HEADLINE — The *Fisherman*, "DFO gives the
green light to more foreign egg imports,"
November 18, 1985.

                    NARRATOR
          The thought of allowing
          Atlantic salmon eggs into
          British Columbia triggered a
          flurry of memos and meetings
          between provincial and federal
          officials.

LETTER — Dave Narver to Bruce Hackett.
November 7, 1986. The words "The introduction
of exotic races of salmonids is probably
the most critical issue ever to face the
maintenance of wild salmonid stocks" lift off
the page and are highlighted.

The policy to bring in Atlantic salmon eggs was labelled by Canadian bureaucrats at the provincial level of government as "an act of biological insanity." The importing of these eggs from Europe was thought to put Canadian salmon stocks at risk of contracting diseases that did not exist in BC waters. (Credit: The Fisherman Publishing Society)

Fisheries　　　　Pêches
and Oceans　　　et Océans

Fisheries - Pacific Region　　Pêches - Région du Pacifique
1090 West Pender Street　　　1090 rue Pender ouest
Vancouver, B.C.　　　　　　　Vancouver (C.B.)
V6E 2P1　　　　　　　　　　　V6E 2P1

February 3, 1987

Mr. B. A. Hackett
Assistant Deputy Minister
Financial Assistance Program
Province of British Columbia
Ministry of Agriculture and Fisheries
Victoria, B.C.
V8W 2Z7

Dear Mr. Hackett:

Thank you for your letter in which you expressed the concern that the proposed Federal-Provincial policy on the importation of live salmonids into British Columbia does not address the need for high quality, domestic Atlantic salmon.

Because, once introduced, an exotic fish disease will, in all likelihood, prove impossible to eradicate, we have placed major emphasis on minimizing the chance of disease accompanying the Atlantic salmon eggs from abroad. For this reason the policy has been to accept eggs from Scotland rather than continental Europe because of Scotland's freedom of VHS and whirling disease. At the same time, we recognize the need for high quality stock which performs well under the conditions of intensive fish culture.

Initially, only ranched or wild Atlantic salmon managed by the Allt Mohr Hatchery in Scotland were certified in compliance with the Canadian Fish Health Protection Regulations. However, two additional facilities, McConnel Salmon and Marine Harvest, in Scotland have now successfully met the certification requirements for cultured (domesticated) fish. Both facilities are operated by successful Atlantic salmon farming companies which plan to export eggs from these stocks into British Columbia in 1987. These stocks originate from a number of wild populations and should contain a reasonable degree of genetic diversity with minimal in-breeding. This being the case, it is unlikely that the quality of the Atlantic salmon stocks available to British Columbia will be a limiting factor in the success of the species on this coast.

000103

Many letters and internal memoranda were exchanged during the years fish farms were planned and introduced to BC's coastal waters. The decision to introduce them was a political one. Many top bureaucrats and scientists were against the move. But elected officials had the final say. Moves were made to try to prevent the introduction of dangerous diseases because the experts knew that once these diseases were introduced, they would be impossible to eradicate. (Credit: Government of Canada)

                    NARRATOR
          One senior manager for the
          province said, "The introduction
          of exotic races of salmonids is
          probably the most critical issue
          ever to face the maintenance of
          wild salmonid stocks."

EXT. NET PEN — DAY

Some Atlantic salmon are dumped into a small
tub.

                    NARRATOR
          And a director general for the
          Department of Fisheries and
          Oceans wrote,

MEMORANDUM — Pat Chamut to B. Hackett.
February 3, 1987. The words "Once introduced
an exotic disease will, in all likelihood,
prove impossible to eradicate" lift off the
page and are highlighted.

                    NARRATOR
          "Once introduced an exotic
          disease will, in all
          likelihood, prove impossible to
          eradicate."

INT. *FISHERMAN* NEWSPAPER OFFICE (1980s) — DAY

Establish. Interview. A young Geoff Meggs
reporting on the perils of fish farms many
years ago.

                    GEOFF MEGGS
          Well, it carries with it the
          threat of disease from the
          Atlantic salmon entering the
          Pacific stocks. It's something
          that we've opposed, that the
          Gillespie inquiry opposed. And
          we, uh, and most biologists
          that have looked at it say that
          the importation of Atlantics
          is like playing Russian
          roulette with our wild stocks.
          And that's precisely what is
          happening. And the more that
          the import of Atlantics that
          goes on, the greater likelihood
          there is of escapes and disease
          problems that we simply won't
          be able to contend with.

INT. HATCHERY — DAY

The water is alive with small fry. A bit of
feed excites them even more.

                    NARRATOR
          The letters and memos
          exchanged made it clear that
          officials were worried about
          the introduction of exotic
          diseases . . .

INT. HATCHERY TANK — DAY

Atlantic salmon swim around aimlessly.

                    NARRATOR
          . . . that had devastating
          impacts on European fish
          populations.

EXT. OCEAN — DAY

Wild fish enter a river.

                    NARRATOR
          But in spite of the risk of
          introducing an exotic disease,

INT. HATCHERY — DAY

A large bank of trays stretches the length of
the room.

                    NARRATOR
          . . . the perceived benefits of
          growing Atlantic salmon seemed
          too enticing to pass up.

                    BRAD HOPE (O.S.)
          There was always danger of
          disease . . .

RESUME — Brad Hope interview.

                    BRAD HOPE
          . . . and the bureaucrats I
          think were absolutely right to
          say let's not bring this stuff
          over and dump it in unless
                    (MORE)

                    BRAD HOPE (CONT'D)
              we are absolutely sure. Let's
              raise the first generation
              on land and let's make sure
              we're coming from disease-free
              stocks.

PHOTO — June Hope is holding a mature chinook salmon by the gill.

                    JUNE HOPE (O.S.)
              I think our decision to bring
              in the Atlantic salmon eggs was
              brought about because I saw how
              easy . . .

INT. HOPE RANCH — MOMENTS LATER

Establish. Interview. June Hope, co-owner of Tidal Rush Farms, with her husband, Brad Hope.

TITLE: June Hope, Former Owner, Tidal Rush Farms.

                    JUNE HOPE
              . . . the Atlantics were to
              handle. Once they were in the
              net pens, when you go up to,
              ah, a raceway of, ah, chinook
              salmon . . .

EXT. SECHELT FARM — DAY

A farmer tosses eggs into a net-covered pen.

JUNE HOPE (O.S.)
. . . that are just in the fry
stage still, they all run away
from you.

RACEWAY — Fry suddenly scoot away.

JUNE HOPE (O.S.)
They just get totally freaked
out, and there's a mad
rush away from where you're
standing.

RESUME — June Hope interview.

JUNE HOPE
You went up to a raceway with
Atlantic salmon in it and
they didn't seem to pay any
attention, at all. And the same
thing happened in net pens. I
saw them, ah, sorting fish, and
handling fish, and they're not
being a problem afterwards. If
we tried to do that with the
chinook salmon, we would run
into all kinds of problems.

INT. FISH FARM PEN — DAY

A shot of chinook salmon. They swim
frantically around.

                    BRAD HOPE (O.S.)
               The Atlantics are just more of
               a domesticated animal.

EXT. FISH FARM — DAY

A farmer tosses some food into the pen.

                    BRAD HOPE (O.S.)
               You can stock them at higher
               density, they like the
               commercial feed, they grow
               fast, they're more disease-
               resistant. I mean, they're
               ideal for farms and the
               Pacifics were not. More net
               pens are more work, and more
               capital and more everything.

                                        FADE TO:

EGGS — They're being poured into a tray.

PHOTO — An aerial view of a Scottish fish farm.

                    JUNE HOPE (O.S.)
               I think that was in 1983 that I
               went to Scotland, and . . .

INT. HATCHERY — DAY

A tray of alevins, just hatched, are wiggling with life.

RESUME — June Hope interview.

> JUNE HOPE
> . . . it took us quite a bit of time to prove that there was a disease-free facility in Scotland and they had to be tested . . .

PHOTO — Atlantic alevins.

> JUNE HOPE
> . . . quite a few times before our government was . . .

PHOTO — A close-up of an eyed salmon egg.

> JUNE HOPE
> . . . satisfied that it was definitely disease-free.

LETTER — Pat Chamut to Bruce Hackett. February 3, 1987. Pan down to the words "to accept eggs from Scotland rather than continental Europe because of Scotland's freedom of VHS and whirling disease." They lift off the page and are highlighted.

> NARRATOR
> After two years of investigation, a decision had been made by Ottawa to "accept eggs from Scotland rather than
> (MORE)

NARRATOR (CONT'D)
continental Europe because of
Scotland's freedom of VHS and
whirling disease."

TIGHT ON EGGS — The developing alevins can be seen through the shell.

BRAD HOPE (O.S.)
I think the original licence
was for either 400,000 or
800,000 eggs.

RESUME — Brad Hope interview.

BRAD HOPE
So we said ok, well, sure,
let's, let's do it and so we
were all prepared, and as I say
my wife went over to Scotland,

PHOTO — June is sitting at a desk doing paperwork.

BRAD HOPE (O.S.)
. . . found the disease-free
hatchery, did all the work,
spent a long time over there
working on it.

RESUME — Brad Hope interview.

BRAD HOPE
Just before we were bringing it in, it was actually a Rockefeller company, I believe, from the US came along and had a lot of money and a lot of authority. And first of all came along and said they'd like to make a deal with you guys. Can we be your partner? So, um, that was fairly early on. And so, and so we, of course, were desperate for money to keep this operation going. So we said yep, we can, and worked with them for a few months, ah . . . And then introduced them to everybody, all through the government, went through all the kinds of things with them. And of course, they were going to help us and what not and, ah, then they said, no, we don't want to do it with you. We want to do it by ourselves, went to the government right away, bang, had all the same licences we did and the government said well, you're going to split your, you'll get half and they'll get half. Which was a little annoying.

LETTER — Gary Hoskins to Dr. Scott. August 27, 1984. The list of quarantine procedures is laid out over three pages.

> NARRATOR
> The government outlined the design and operation requirements for a quarantine facility in letters to the two importing farmer groups. On paper, the procedures seemed strict and comprehensive, and both groups scrambled to meet the criteria.

PHOTO — Pacifica Aqua Foods, Port Alice Hatchery.

EXT. VANCOUVER ISLAND (MAP) — DAY

The two locations where the Atlantic eggs landed fade up. One near Port Alice. The other near the Kokish River.

> NARRATOR
> But no matter how good the procedures, it was a scary time for the civil servants who had spent their lives protecting wild salmon populations.

PHOTO — Tidal Rush Farm's yellow boat.

PHOTO — A farmhand walks along a net pen rim with a net.

                    NARRATOR
          And it became even scarier
          when the Americans saw the
          importation of . . .

MEMORANDUM — W. Doubleday to Pat Chamut.
"Trade Issue with USA — B.C. Atlantic Salmon
Policy." July 30, 1992. The subject of the
memo is highlighted.

                    NARRATOR
          . . . Scottish eggs to British
          Columbia as a business
          opportunity and turned Atlantic
          salmon eggs into a free-trade
          challenge.

RESUME — Brad Hope interview.

                    BRAD HOPE
          The Americans had said, who
          then started into salmon
          farming, there were some salmon
          farms at Anacortes. They'd kind
          of gotten wind of what was
          happening in Canada. And they
          started doing it, and so they
          of course, going through the
          American government, applied
          and they said, oh yeah, do
          whatever you want. So they were
          bringing in Atlantic salmon,
          right over across the way. You
          could see it from Victoria
                      (MORE)

                    BRAD HOPE (CONT'D)
          virtually, and ah, there was
          no disease control. So our
          argument was let us bring them
          in and let us supply other
          places and, um, and that'll
          make a lot of sense. We'll have
          disease-free stock coming in.
          We'll quarantine it, we'll make
          sure it's clean when it goes
          into the ocean and we won't
          have to deal with any place,
          even the US if we can get them
          to cooperate.

INT. HATCHERY TANK — DAY

Atlantic salmon are reared in a hatchery.

                    NARRATOR
          But the Pacifica Aqua Foods
          plan fell on deaf ears, and
          within a year, Atlantic salmon
          eggs were flowing from several
          sources.

WORLD MAP — Animated planes act like arrows
to show where the eggs were coming from and
landing in BC.

                    NARRATOR
          Scotland, Iceland, Ireland,
          Eastern Canada and Washington
          State.

PHOTO — Workers are taking care of some brood stock.

> NARRATOR
> And as it turns out, the concerns expressed by Canadian officials may not have mattered at all.

PHOTO — The Washington State hatchery.

> NARRATOR
> Washington State had been experimenting with Atlantic salmon at least five years before British Columbia allowed their importation . . .

INT. OCEAN — DAY

A school of herring fill the frame.

> NARRATOR
> . . . and the fact that herring populations began to decline on the southwest side of Vancouver Island . . .

REUSME — Animated map. This time Puget Sound goes red as well.

> NARRATOR
> . . . before areas near BC's early farms suggests that a deadly new disease had already
> (MORE)

                    NARRATOR (CONT'D)
          taken hold in the environment
          and was making its way north
          from Puget Sound.

HEADLINE — The *Fisherman*, "All's not well
down on the farm," August 16, 1985.

                    NARRATOR
          And then, a few years later,
          when Atlantic salmon were
          imported into BC,

HEADLINE — The *Fisherman*, "Danger at the
farm," April 18, 1986.

THE WORDS — "eggs certified as disease free
at the Scottish hatchery were responsible
for an outbreak of furunculosis" lift off the
page and are highlighted.

                    NARRATOR
          . . . a report surfaced that
          "eggs certified as disease free
          at the Scottish hatchery were
          responsible for an outbreak of
          furunculosis" at 28 fish farms
          in Norway.

EXT. RIVER — DAY

Someone is looking at a pile of pre-spawn
mortalities of chum. He splits the belly
of one to reveal that it is full of sacs of
milt, meaning it died before spawning.

WIDER — Dead chum are everywhere in the stream.

ON A SINGLE FISH — It has a big bulge on its side, a typical sign of furunculosis. Someone is poking at it with a knife.

> ALEXANDRA MORTON (O.S.)
> So one year we got this big bacterial outbreak in the hatchery. Um, 28 percent of the fish died. And ah, Fisheries and Oceans Canada came and took a look. They said it's furunculosis.

RESUME — Morton interview.

> ALEXANDRA MORTON
> Oh! Furunculosis. We'd never heard of that. And they gave us a drug and we injected the fish and they healed up miraculously and everything was good.

EXT. ALEXANDRA'S DOCK — DAY

Her boat is moored.

> ALEXANDRA MORTON (O.S.)
> People from the fish farming industry were living in the community . . .

EXT. SMALL FISH FARM — DAY

A farmer pulls his dinghy along a line secured to shore.

>   ALEXANDRA MORTON (O.S.)
>   . . . and they're like, "Oh, we
>   had furunculosis back in June
>   on the farms." And our coho
>   are coming back in September,
>   October.

EXT. SMALL FARM NET PEN — DAY

A farmer sprinkles fine fish food into the fry pen.

>   ALEXANDRA MORTON (O.S.)
>   So we knew that they had
>   furunculosis first.

TIGHT — The fry react to the food and boil at the surface of the small pen.

>   ALEXANDRA MORTON (O.S.)
>   And a couple of years later, I
>   was at the gas dock gassing up
>   my boat, and this kid from a
>   farm goes,

A BIT WIDER — The fry continue to feed.

ALEXANDRA MORTON (O.S.)
. . . "Hey, we have a triple dose of furunculosis." I was like, "Ooh, a triple dose of furunculosis.

AND VERY TIGHT — on the fry feeding away.

ALEXANDRA MORTON (O.S.)
What is that?!" So, I began writing letters.

INT. HATCHERY — DAY

A worker opens the cover. Alevins squirm in the tray.

ALEXANDRA MORTON (O.S.)
It turned out to be a, a strain of the disease that was resistant to three drugs. It was triple-antibiotic-resistant furunculosis.

EXT. HATCHERY — DAY

A brood stock pond is full of mature fish.

ALEXANDRA MORTON (O.S.)
I wrote letters to DFO. And I wrote letters to the province.

EXT. POND — DAY

Smolts swim in a circle in a small pen.

ALEXANDRA MORTON (O.S.)
And I wrote letters to the fish
farmers, and they didn't get
it.

LETTER — Alexandra Morton to Pat Chamut.
December 9, 1993. Part of the letter is
highlighted.

ALEXANDRA MORTON (O.S.)
And so I thought ok, ok. Let me
try this again. And I . . . I
kept thinking, if I just could
line up the words in the right
order, they would get it.

RESUME — Morton interview.

ALEXANDRA MORTON
And I was the only biologist
in my community, and so I felt
like it was my responsibility
to be the one that brought the
community concerns forward.

LETTER — Pat Chamut to Alexandra Morton. May
10, 1993.

NARRATOR
The Department of Fisheries and
Oceans did some testing, and
in a letter sent to Morton they
confirmed that . . .

PAN TO — the words "I agree," then to
"strain," then "multiple drug resistance,"
then to "new occurrence in B.C.," with each
lighting up.

> NARRATOR
> . . . "Isolation of a strain"
> that is resistant to three
> drugs "is a new occurrence in
> BC."

MEETING MINUTES — Front page of the
minutes with a Province of British Columbia
letterhead. December 9, 1993.

> NARRATOR
> But a joint meeting between
> federal and provincial
> ministries concluded that
> the . . .

PANNING DOWN — The words "tested," "Scott
Cove" and "low-dose" are highlighted.

> NARRATOR
> . . . strain of furunculosis on
> the Broughton fish farms was
> different from the outbreak at
> the Scott Cove hatchery.

INT. SCOTT COVE HATCHERY — DAY

Billy Proctor and an assistant are loading
eggs into a tray.

Almost an entire run of chum salmon died in Kakushoish Creek on Denny Island on BC's central coast. This photo was taken on September 10, 2012, when researchers were sampling the fish for diseases that would cause them to die before spawning. This phenomenon is called pre-spawn mortality and has been occurring in many rivers and streams along the BC coast since the early 1980s. (Photo credit: Jody Eriksson)

This coho salmon was found bloated and was bleeding into its abdomen. It also died before spawning, which is a typical symptom of VHS. (Photo credit: Jody Eriksson)

NARRATOR
But how could a hatchery in the middle of nowhere suddenly have a strain of bacteria resistant to drugs that just happen to be used on fish farms?

BACK TO THE MINUTES — On page 3, the words "triple sulpha antibiotics" are highlighted.

NARRATOR
And the solution was to treat the outbreak with triple sulpha-antibiotics, which sounded a lot like what the young fish farmer called triple dose furunculosis.

RESUME — Morton interview.

ALEXANDRA MORTON
I was so naïve. I really thought the government was working for the people, and they would address this if I could just make it clear to them that this was serious and it was new and, and, and something was going on here that wasn't right.

FADE OUT:

FADE IN:

EXT. CAMPBELL RIVER PODIUM — DAY

Rich Hagensen and his wife are on the stage. He sings "The Get Out Migration."

></br>RICH HAGENSEN (O.S.)
> On the BC coast, there's a growing opposition . . .

EXT. CAMPBELL RIVER — DAY

The protesters are walking along a street.

SUPERSCRIPT: Get Out Migration Protest — Day Five.

> RICH HAGENSEN (O.S.)
> . . . from citizens and First Nations, to stop those big Norwegian fish farm corporations,

ANOTHER STREET — The protesters turn a corner.

> RICH HAGENSEN (O.S.)
> . . . and save the wild salmon migrations.

EXT. FIELD — MOMENTS LATER

The protesters are carrying a Norwegian flag and a sign reading "Marine Harmfest."

Fish farm protesters dump dead Atlantic salmon at the front door of Marine Harvest in Campbell River on April 28, 2010. (Photo credit: John Preston)

Get Out Migration protesters arrive at the Parliament building in Victoria, BC, on May 8, 2010. It was one of the largest protests to ever descend on the BC legislature lawn. (Photo credit: John Preston)

Alexandra Morton surveys the thousands of protesters who joined her on the last day of the Get Out Migration protest while Billy Proctor talks about the questionable methods used to locate fish farms in the Broughton Archipelago. (Photo credit: Jack Cooley)

EXT. FISH FARM OFFICE BUILDING — DAY

The protesters drop Atlantic salmon on the doorstep.

>   RICH HAGENSEN
> The Minister of Environment says the wild salmon are not harmed while swimming past fish pens of shit and lice,

A PROTESTER IN A HAZMAT SUIT — They add to the pile of Atlantic salmon.

>   RICH HAGENSEN (O.S.)
> . . . and the DFO says there's no evidence that Slice,

BACK TO THE PODIUM — Starting on his guitar, pans up to Rich singing.

>   RICH HAGENSEN
> . . . that pesticide don't kill the fish farm lice, real nice! But they're wrong,

EXT. FISH FARM OFFICE BUILDING — DAY

Alexandra is talking with a reporter.

WIDER — Lots of media around Alexandra.

>   RICH HAGENSEN (O.S.)
>   . . . says Alexandra Morton's
>   probe. And it's echoed by
>   voices round the globe.

ON ALEXANDRA — She's now at a podium ready to speak.

>   RICH HAGENSEN (O.S.)
>   Open net fish farms are
>   pollutin' up the seas, and this
>   we can say without a doubt.

BACK TO RICH

>   RICH HAGENSEN
>   They're killing off the wild
>   Pacific salmon, so get 'em up
>   on shore or get out.

                                        FADE OUT:

FADE IN:

EXT. FISH FARM (AERIAL) — DAY

Establish. A large fish farm sitting in a channel.

>   JOHN VOLPE (O.S.)
>   We're looking at the effects of
>   the fish farms acting as fish
>   aggregating devices.

INT. FISH FARM (UNDERWATER) — DAY

A mixture of fish are swimming in the water inside the pen. Herring and perch are mingling with the farmed salmon.

> JOHN VOLPE (O.S.)
> So you have these farms in essentially the open ocean, ah, producing tremendous amounts of nutrients both in the form of uneaten feed on the farms,

OUTSIDE THE PENS — Small fish school around the pens.

> JOHN VOLPE (O.S.)
> . . . but also the fish feces, which attract large numbers of small fish. The small fish, of course,

RESUME — Volpe interview.

> JOHN VOLPE
> . . . when they congregate in high densities will attract large fish.

INT. CHINOOK FARM NET PEN — DAY

Lots of large herring and perch in with the farmed salmon.

SUPERSCRIPT: Inside an open net pen.

>   JOHN VOLPE (O.S.)
>   And so you've got, presumably,
>   or, or potentially, a complete
>   reconfiguration of the
>   ecosystem.

EXT. FISH FARM (1980s) — DAY

A worker is feeding the fish in a pen, tossing handfuls of feed into the water.

>   ALEXANDRA MORTON (O.S.)
>   Fish farms are basically
>   chumming the water. I don't
>   know if you've ever seen the
>   movies of sharks where they
>   throw blood and food in the
>   water and the sharks are
>   attracted to it.

PANNING DOWN — The farm fish come up to take the food in the pen.

>   ALEXANDRA MORTON (O.S.)
>   Well, everything likes to eat
>   fish food.

RESUME — Morton interview.

>   ALEXANDRA MORTON
>   The seagulls have learned how
>   to pick it out of the air when
>   the blowers are going.

EXT. ROUND PEN FISH FARM — DAY

Salmon jump inside the pens.

> ALEXANDRA MORTON (O.S.)
> It creates an oil slick. It's a valuable source of oil to all the fish in the area.

OUTSIDE THE PEN — Salmon fry school around looking for food.

> ALEXANDRA MORTON (O.S.)
> Everything wants oil, and so they're going to be attracted to this food and they're going to eat it.

LOW ANGLE — Looking back into the pens.

> ALEXANDRA MORTON (O.S.)
> When they blow it out of the feeders, there's this dust of it. When the sunlight is just right you can see it and it settles on the water.

ON THE WATER — Thousands of fry are feeding on something just outside the pens.

> ALEXANDRA MORTON (O.S.)
> And there's actually this little oil sheen that, that spreads out from every farm.

TRACKING ACROSS — The side of a farm.

>   ALEXANDRA MORTON (O.S.)
>   And so everything's feeding
>   on it.

BACK TO THE FRY — They're feeding next to the farm.

And we fade to . . .

CLOSE UP — A red scoop goes into a bucket of large fish food pellets.

EXT. FISH FARM (ARCHIVAL) — DAY

The worker tosses the feed into a large pen. Farm fish rise to the feed. And again. And again.

FOOD PELLETS — The pellets drift downward through the pen among the fish.

>   JOHN VOLPE (O.S.)
>   In BC there seems to be some,
>   ah, asymmetry with the rest
>   of the salmon-producing world.
>   Everybody looks at, you know,
>   BC numbers are really good.

INT. FISH FARM PEN (UNDERWATER) — DAY

A worker is standing on the edge of the pen.

                    JOHN VOLPE (O.S.)
          Right? Is it because they are
          just cooking the numbers?

MOVING LEFT — The fish take the pellets.

                    JOHN VOLPE (O.S.)
          Or are they really that good?
          Well, if they're really that
          good, one reason might be
          because the diet is being
          supplemented by these pit-
          lamped . . .

RESUME — Volpe interview.

                    JOHN VOLPE
          . . . farms that are bringing
          in supplemental feed in the
          form of, you know, all these
          wild critters.

HEADLINE — The *Fisherman*, "Young salmon, herring are farm food," April 21, 1989.

                    NARRATOR
          As it turns out, this suspicion
          by today's fish farm critics
          was reported years ago,

ROB RUSSELL MEMORANDUM — The words "Interviews with on-site staff confirm that farmed salmon consume these herring to the point where additional feeding is unnecessary at some times" are highlighted.

                    NARRATOR
          . . . in 1988, by a DFO habitat
          biologist in an internal memo.
          The DFO staffer observed
          that . . .

INT. OCEAN, NEAR FISH FARM PEN — DAY

Schools of very small herring are milling
around under a farm.

                    NARRATOR
          . . . "Juvenile herring
          schooling around the pens
          are consumed to the point
          additional feeding is
          unnecessary."

INT. OCEAN — The current pushes the sea life
around.

INT. OCEAN — DAY

Small fish are schooling under a wharf.

WIDER — There are thousands of fish.

                    BRAD HOPE (O.S.)
          We had a couple of sites that
          were, where there was good
          tides and . . . I remember, ah,
          small herring coming through.
          Boy, the fish would just thrash
                         (MORE)

                    BRAD HOPE (O.S.) (CONT'D)
               and then you would go and
               feed and there wasn't a
               movement, they were just not
               interested . . .

RESUME — Brad Hope interview.

                    BRAD HOPE
               . . . in what we were giving
               them, and it would be a day or
               two and then they'd be back. We
               saw that many times.

PHOTO — Hidden Basin Farm with a float plane docked at it.

                    BRAD HOPE (O.S.)
               We actually got ourselves in
               trouble with the bank because
               ah,

PHOTO — New fry runways are being placed next to the farm support building.

INT. TANK — DAY

Atlantic salmon swim around in the tank.

                    BRAD HOPE (O.S.)
               . . . because you have to feed
               them. And when you've got a
               growing mass, a mass that's
               growing, doubling in some cases
               every three weeks,

PHOTO — A young Brad stands with his arms crossed, looking out at the water.

> BRAD HOPE (O.S.)
> . . . the outlay of cash was equally doubling every three weeks.

EXT. ANOTHER FARM — DAY

A farmer is pulling handfuls of feed out of a five-gallon pail.

> NARRATOR
> So with food costs so critical to the operation of a salmon farm, any free food would not only be welcome,

Another farmer is making notes.

> NARRATOR
> . . . but would be a crucial part of the farming operation.

EXT. LEGISLATURE (MAY 8, 2010) — DAY

Billy Proctor stands at the microphone in front of thousands of protesters.

> BILLY PROCTOR
> They called us, all us old farts, in and told us that we had to tell 'em where the
> (MORE)

                    BILLY PROCTOR (CONT'D)
          migration routes in the wild
          salmon were. This was about
          18 or 20 years ago, so all of
          us fishermen in the Broughton,
          and around Alert Bay, Sointula,
          gathered in Alert Bay, and we
          drew up a map and DFO put on
          them green zones, yellow zones
          and red zones. And now out of
          the 17 farms in the Broughton,
          there's 11 in the red zones
          where there was not supposed to
          be any.

EXT. EARLY FARM — DAY

A seiner has its net laid out next to some
pens.

                    NARRATOR
          But it isn't just the placement
          of the farms or the chumming of
          the water the farmers could use
          to attract small fish into the
          pens. There is one other method
          that is seen as necessary to
          keep the farm fish eating all
          year round.

                                        FADE OUT:

FADE IN:

EXT. FISH FARM — NIGHT

An eerie shot of a fish farm with its pens lit up.

TRAVELLING PAST — THE FISH FARM

A lone fish farm worker walking along a wharf with a flashlight pointed down. A fish can be seen flopping through the mesh. It's eerie.

FISH — They swim past nets in the dark. Then small fish scurry through the light.

          ALEXANDRA MORTON (O.S.)
    Their pens are full of wild
    food, and they will not let
    anybody observe at night on
    their pens to see if those fish
    are being eaten, and they will
    not let observers . . .

RESUME — Morton interview.

          ALEXANDRA MORTON
    . . . at the processing plants
    to look in the stomachs of
    these fish.

BC's fish farms are lit up at night. The farmers say it is so their fish will grow faster. However, observers now know that part of the reason is that the lights attract juvenile wild fish and bait fish into the pens to be eaten. This unregulated harvesting of wild fish has had a large impact on wild fish populations. It also means that wild fish and farmed fish are in much closer proximity to each other, allowing a greater exchange of pathogens. (Photo credit: Juggernaut Pictures)

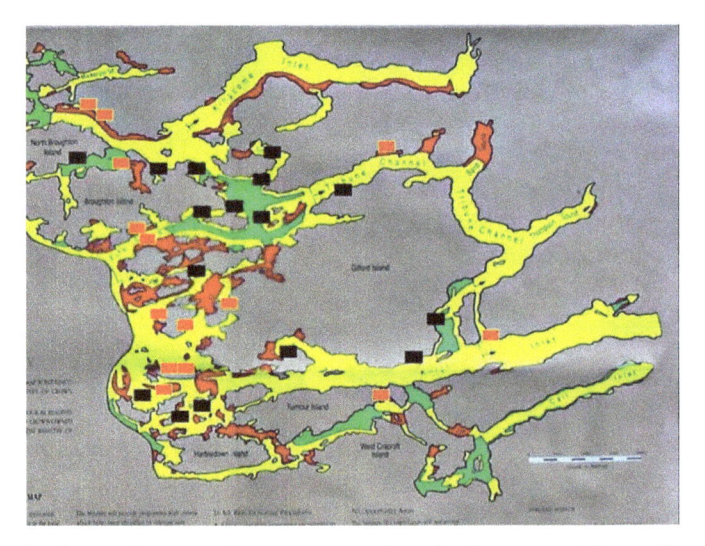

This diagram, known as the CRIS map, outlines the three zones delineated by public consultation with locals in the Broughton Archipelago. The green zones are considered to have the least impact on wild fish populations, the yellow zones a moderate impact and the red zones the greatest impact. It turns out that most of the farms were located in the "productive" red zones so fish farms could benefit from the productivity of the food web in those areas. (Credit: Province of British Columbia)

INT. BILLY PROCTOR'S LIVING ROOM — DAY

Establish. Interview. Billy Proctor, long-time resident of the Broughton Archipelago in Echo Bay, BC.

TITLE: Billy Proctor, Commercial Fisherman, Echo Bay, BC.

          BILLY PROCTOR
Like on the farm in the Birdwoods, for instance, they have thirty-six 1000 watt bulbs . . .

EXT. OCEAN — NIGHT

A fish farm is lit up with intense lights. It's weird and spooky.

          BILLY PROCTOR (O.S.)
. . . underwater going all night. And I've gone out there and ran around the farm with my boat.

CLOSER — A fish farm's lights light up the farm building.

          BILLY PROCTOR (O.S.)
I've got a good sounder on there, and you run around the boat and there's so much fish outside the farm that it . . .

EXT. FISH FARM — NIGHT

A school of anchovies swim toward the light.

> BILLY PROCTOR (O.S.)
> . . . blocks the sounder that you can't pick up the bottom through it. There's that much fish attracted to these lights.

> MARTY KRKOSEK (O.S.)
> A consequence of having these bright lights on at night is it attracts a lot of things in the sea towards them. We know this.

SUB-HEADLINE — "The use of vapour lights."

THE WORDS — "The use of vapour lights is the death knell and swan song of our herring" lift off the page and are highlighted.

HEADLINE — The *Fisherman*, "Partial Ban Ordered on Pit-Lamping Herring," August 6, 1965.

INT. SALMON RESEARCH STATION LIBRARY — DAY

RESUME — Krkosek interview.

> MARTY KRKOSEK
> Pit-lamping for herring was, you know, outlawed decades ago because it was too effective.

The local herring population around Tidal Rush Farms disappeared after just a few years of the fish farm being in operation at Nelson Island's Hidden Basin, located on British Columbia's Sunshine Coast. The farmers had discovered that herring would swim into the farm and be eaten. When the herring stocks were depleted, the farmers devised a plan, with the help and blessing of the government, to import surplus herring eggs from other locations to hatch them out in Hidden Basin to try to restart the local run. The experiment was unsuccessful, but the project team could not pinpoint why. (Photo credit: Doug Hay)

During large spawn events, many herring eggs are dislodged from spawning surfaces by wave action, especially during storms. The eggs are washed up on shore and form a wide band of eggs and debris. Many are refloated during a high tide and presumably hatch out. But many perish. These eggs were the target of transplants to try to restart herring populations that had disappeared from other areas, usually due to over-harvesting or fish farm activities. (Photo credit: Doug Hay)

Wind-drift herring eggs were scooped up in a pail and
loaded onto a boat for transplant. (Photo credit: Doug Hay)

INT. OCEAN — EVENING

From under a farm, looking up. Small fish mingle around the net.

> NARRATOR
> But it isn't just herring that were being drawn into the pens and being eaten.

BACK TO MEMORANDUM — The words "Interviews with processing plant staff have also confirmed the presence of salmon fry in the stomach of farmed fish" pop up.

> NARRATOR
> "Interviews with processing plant staff have also confirmed the presence of salmon fry in the stomach of farmed fish."

INT. VANCOUVER CONDO — AFTERNOON

Establish. Interview.

TITLE: Don Staniford, Anti-Salmon Farming Activist.

> DON STANIFORD
> The biggest fear is that you've got huge runs of wild salmon. Um, and if you're prepared to sacrifice that at the altar of salmon farming,

EXT. ADAM'S RIVER — DAY

A large school of sockeye salmon fill a river.

>   DON STANIFORD (O.S.)
>   . . . then I, I think that's
>   stupid. That's reckless abandon
>   from the provincial and federal
>   government here . . .

INT. ADAM'S RIVER — MOMENTS LATER

Bright red sockeye fill the frame.

>   DON STANIFORD (O.S.)
>   . . . in Canada and British
>   Columbia.

FROM ABOVE — A pool is crammed with sockeye.

>   DON STANIFORD (O.S.)
>   And ah, if that's the situation
>   where wild salmon are going to
>   be sold down the river, then at
>   least the public should be, uh,
>   told about that.

FROM SHORE — More sockeye in a shallow stretch of the river.

>   DON STANIFORD (O.S.)
>   There should be a public
>   debate. This should be a public
>   decision.

                    NARRATOR
          "Sacrifice" sounds like a harsh
          word.

EXT. LARGE CORPORATE FARM — DAY

This farm is massive.

                    NARRATOR
          But the unregulated grazing on
          wild juvenile fish is certainly
          sacrificing the economic
          potential of the mature fish
          they represent. And it appears
          that the sacrifice may have
          been swift.

EXT. FISH FARM (1980s) — DAY

A small farm on the Sunshine Coast.

                    NARRATOR
          In the early 1980s, farmers
          and DFO personnel tried
          to re-establish herring in
          areas where they suddenly
          disappeared.

                                        FADE OUT:

FADE IN:

A roll of film ramps up. A jerky shot of a camera assistant is in frame.

SUPERSCRIPT — Dick Harvey audio recording, 1982.

                    BRAD HOPE (V.O.)
          Is the recorder on now, Dick?

                    CAMERA ASSISTANT (V.O.)
          Take 1.

EXT. HERRING POND — DAY

Two guys are crouched on boom sticks that surround a pen.

                    BRAD HOPE (V.O.)
          We have a family operation
          which is primarily a salmon
          farm, ah, in Hidden Basin on
          Nelson Island.

PHOTO — The yellow Tidal Rush boat at a launch site.

                    BRAD HOPE (V.O.)
          This year we've been doing
          a herring transplant project
          which is taking wind-drift
          herring roe, that's herring
          eggs that are . . .

PHOTO — A shot of wind-drift herring roe along a beach.

> BRAD HOPE (V.O.)
> . . . washed up on the beaches every year through storms, taking that roe which would normally die, which always dies on the, on the beaches as a result of being blown up.

PHOTO — Someone is scooping up some herring eggs that are lying on seaweed.

> BRAD HOPE (V.O.)
> Bringing it back to Hidden Basin, um, the site of our farm, and ah, distributing it around the basin. Will they home back to this site?

PHOTO — Two people are gathering eggs using a herring skiff.

> BRAD HOPE (V.O.)
> If the answer to that is yes they will, then the possibility of enhancing herring on the Pacific coast, ah, looks extremely promising . . .

PHOTO — Full frame of herring eggs on the beach.

                    BRAD HOPE (V.O.)
               . . . in many different areas
               where, ah, herring once existed
               and where they no longer exist.
               I don't know whether that was
               worth adding.

PHOTO — A man is lying, partially covered, in
the herring eggs.

                    DICK HARVEY (V.O.)
               Starting with Bob here now. Bob
               McIlwaine, take 2.

PHOTO — Wind-drift herring eggs cover a large
beach.

                    BOB MCILWAINE (V.O.)
               Ah, Chief of Fisheries
               Development Division, the
               Department of Fisheries and
               Oceans.

PHOTO — Two workers measuring the beach
herring eggs.

                    BOB MCILWAINE (V.O.)
               Now you were wondering, Dick,
               how this project came about to
               salvage herring . . .

PHOTO — A farmer standing in the beach eggs.

                    BOB MCILWAINE (V.O.)
               . . . eggs from the beaches in
               an attempt to grow them out in
               Hidden Basin.

PHOTO — Someone pushing a pile of eggs into a
metal pail.

                    BOB MCILWAINE (V.O.)
               There is a tremendous loss
               of these eggs every year
               from storms and other, other
               factors. And the potential, the
               possibility existed to take the
               eggs,

PHOTO — Looking down into a school of herring
near the surface.

                    BOB MCILWAINE (V.O.)
               . . . move them to an area
               and grow them out, in perhaps
               inlets, where there used to
               be spawning, ah, spawning
               stocks . . .

INT. OCEAN — DAY

A few herring are spawning in an eelgrass
bed.

                    BOB MCILWAINE (V.O.)
               . . . and they no longer exist
               in these areas.

NARRATOR
It's not really clear if
the farmers had overgrazed
the herring in their area
or whether the cause of the
disappearance was from a new
disease spreading through the
environment. But there was no
turning back, and the gamble
was underway that fish farms
would provide the jobs needed
to boost the BC economy.

EXT. MODERN FISH FARM — DAY

It's a calm day. Quiet with only one worker in sight.

NARRATOR
So 30 years later, has the
aquaculture industry created
the jobs that were so alluring
when the industry first began?

RESUME — Morton interview.

ALEXANDRA MORTON
Initially there were jobs.
There was more jobs than my
community could fill. So people
moved in. People moved in with
families. It was great.

EXT. WEST COAST — DAY

An aerial view of a large fish farm.

> ALEXANDRA MORTON (O.S.)
> But then the companies got bigger and got more and more mechanized.

EXT. MODERN LARGE FARM — DAY

No workers are on site.

> ALEXANDRA MORTON (O.S.)
> Ah, with fewer and fewer people on the farms, and now nobody in my community works on the fish farms. Um. Nobody in many of the communities works on the fish farms, which I . . .

EXT. ANOTHER FARM — DAY

No workers on site again.

> ALEXANDRA MORTON (O.S.)
> . . . I tend to think is a plan on the fish farms' part, because while fish farmers were living in Echo Bay, we were friends and, ah, we learned a lot about the fish farms. I mean, every time they opened their mouths we learned something we didn't know.

EXT. YET ANOTHER MODERN FARM — DAY

No workers in sight.

>   DARREN BLANEY (O.S.)
>   That's really not much jobs
>   available . . .

EXT. BEACH — DAY

Establish. Interview.

TITLE: Darren Blaney, Former Chief Homalco First Nation.

>   DARREN BLANEY
>   . . . in fish farms. The one in
>   Church House, there was only
>   one person working there and
>   that was my cousin, and he was
>   making barely above minimum
>   wage. He certainly couldn't
>   support himself on it.

EXT. CHANNEL — DAY

A large farm is tucked against the far side.

>   RAFE MAIR (O.S.)
>   If the fish farms and
>   other things provided huge
>   employment, you know,

CLOSER — There appear to be no workers on this farm.

                    RAFE MAIR (O.S.)
            . . . if thousands of people
            relied upon it, well, that
            wouldn't make it right but at
            least it would give an . . .

EVEN CLOSER — Finally we see one worker
behind the beat-up netting covering the pens

                    RAFE MAIR (O.S.)
            . . . argument that you'd have
            to deal with. But these, ah,
            fish farms, much like the, ah,
            the power plants, the private
            power plants, once they're,
            they are built have very, very
            little employment at all, a
            couple of watchmen.

INT. RAFE'S CONDO — DAY

Establish. Interview. Rafe Mair is a retired
politician and political commentator.

                    RAFE MAIR
            All I'm saying is that one
            of the justifications can
            be employment. And if that
            justification isn't there,
            then what you're doing simply
            is taking bags of gold from
            British Columbia and putting it
            in, putting them in Oslo. And
            it makes absolutely no sense.

GRAPH — This graph compares the sport, commercial and aquaculture sectors in terms of employment.

             NARRATOR
     A study by Stats Canada
     confirms that aquaculture has
     not generated the employment
     promised and is the lowest
     employment generator in the
     fish-exploiting industries.

                              FADE OUT:

FADE IN:

EXT. ISLAND HIGHWAY — DAY

The protesters are making their way up a hill. Holly Arntzen and Kevin Wright sing "I Am the Future."

EXT. DUNCAN CITY LIMIT — DAY

Establish. A sign welcomes visitors to the town.

EXT. DUNCAN STREET — DAY

The protesters cross the street.

             HOLLY ARNTZEN (V.O.)
     When you think of the distant
     future,

Holly Arntzen and Kevin Wright from the ARTist Response Team perform at the Get Out Migration rally in Duncan, BC, on May 6, 2010. (Photo credit: Scott Renyard)

Alexandra Morton leads protesters down a street in Duncan, BC, on May 6, 2010. (Photo credit: John Preston)

EXT. ANOTHER DUNCAN STREET — DAY

Lots of people are waiting for the protesters. They line the street.

>   HOLLY ARNTZEN (V.O.)
>   . . . do you see anything at
>   all? Will the planet keep on
>   spinning?

EXT. CROWDED STREET — DAY

The protesters work their way through the crowd.

>   HOLLY ARNTZEN (V.O.)
>   Or will it fall?

EXT. ANOTHER STREET — MOMENTS LATER

Alexandra leads the crowd.

>   HOLLY ARNTZEN (V.O.)
>   Looking at the constellations
>   that come back every night,

SUPERSCRIPT: Get Out Migration Protest — Day 14

>   HOLLY ARNTZEN (V.O.)
>   . . . fourteen billion years of
>   evolution in sight.

EXT. DUNCAN SQUARE — MOMENTS LATER

They enter the square. Alexandra gets to the podium.

> KEVIN WRIGHT (V.O.)
> Let's do this! I,

> HOLLY ARNTZEN (V.O.)
> I, I am,

Alexandra gets up on the stage.

> HOLLY ARNTZEN (V.O.)
> . . . I am the future, I'm the new, ew, ew. We, we are,

REVERSING — The crowd is gathering around the stage.

> HOLLY ARNTZEN (V.O.)
> . . . we are the future and we're counting on you.

ON A DRUM

> WILLIAM ROUTLEY (O.S.)
> I remember as a little kid standing on the, on the wharf down there in Cowichan Bay and seeing the entire bay just teeming with fish.

EXT. DUNCAN SQUARE PODIUM — DAY

Alexandra smiles and claps. The crowd laughs and cheers.

ON WILLIAM ROUTLEY

TITLE: William Routley, MLA, Cowichan Valley, BC.

                    WILLIAM ROUTLEY
          Fish jumping all over the bay
          and the river just teeming
          with salmon, and I know our, I
          see our First Nations brothers
          and sisters nodding their
          heads. They remember what
          it used to be like, and any
          of you that have been round
          and long in the tooth like
          me knows that, that's true.
          And it's unacceptable what's
          happening. We have to bring
          about change. Thank you, ladies
          and gentlemen, for being here
          and for caring about salmon and
          working towards change. Thank
          you very much.

And he exits the stage.

                                        FADE OUT:

FADE IN:

EXT. OCEAN — EVENING

A remarkably empty channel. It looks pristine.

EXT. ECHO BAY RESEARCH STATION DOCK — EVENING

The dock at the Echo Bay research station sits peacefully under a partially dark sky.

SUPERSCRIPT: Summer 2001

ALEXANDRA — She's standing onshore. Her hair is blowing in the wind.

> ALEXANDRA MORTON (O.S.)
> In 2001, this neighbour of mine,

RESUME — Morton interview. It's a stormy day outside. The wind is howling.

> ALEXANDRA MORTON
> . . . Chris Bennett, has a beautiful little lodge he built by hand.

EXT. CHRIS BENNETT'S HOUSE — DAY

Chris tests a fishing rod.

                    ALEXANDRA MORTON (O.S.)
                He's an amazing salmon
                fisherman, and he comes to me
                with this bucket.

ON A BUCKET — Two fry are swimming around in a white pail.

                    ALEXANDRA MORTON (O.S.)
                And he's got a little chum
                salmon and a little pink salmon
                and the cords on his neck are
                standing out. And . . .

ON A HAND — A little pink fry is in Chris's hand and there's a sea louse on the fry.

                    ALEXANDRA MORTON (O.S.)
                . . . he's like, "Are these sea
                lice on these fish?"

ON TWO FRY — The fry are dead and have sea lice on them.

                    ALEXANDRA MORTON (O.S.)
                'Cause you could see they
                were like bristling with
                these little hairs. And I was
                like, "Boy I don't know. They
                don't . . . you know . . .

INT. ALEXANDRA'S ECHO BAY LAB — DAY

Alexandra is looking at samples under a microscope.

> ALEXANDRA MORTON (O.S.)
> They don't really look like sea lice." "Well, so my guests from Scotland, you know, that's why they're here fishing with me,

RESUME — Morton interview.

> ALEXANDRA MORTON
> . . . because they left Scotland 'cause there's no more sea trout and Atlantic salmon there because the sea lice from the fish farms killed them off!"

A PINK FRY — The fry is in a bag and has sea lice on it.

> ALEXANDRA MORTON (O.S.)
> So I took these little fish and I started to look on the Internet, and indeed they were sea lice.

INT. AQUARIUM — DAY

A small fry with lice on it struggles on the bottom of the aquarium.

> ALEXANDRA MORTON (O.S.)
> Not only is it going on here, but it's going on in Norway. I was,

PHOTO — A pink fry covered with sea lice in a small photarium.

>   ALEXANDRA MORTON (O.S.)
> . . . I was corresponding, ah, with Norwegian scientists right from the beginning. And they kept asking me, "Do you have the sea lice plague yet?"

RESUME — Morton interview.

>   ALEXANDRA MORTON
> And ah, I would be like, "Nope, no. No, no."

RESUME — Mair interview.

>   RAFE MAIR
> I met a delightful fellow named Paddy Gargan, Dr. Patrick Gargan, who is a . . . a fish biologist from County Galway. And he came out to verify Alexandra Morton's findings some years ago.

                    JUMP CUT:

>   RAFE MAIR
> So I went down there and started talking to them and one fellow said, he said, "Can't you fuckin' well read up in Canada? Can't you see
>   (MORE)

                    RAFE MAIR (CONT'D)
          what's happening in Norway, in
          Scotland and here in Ireland?
          Can't you fuckin' well read?"
          And I had to turn around and
          say, evidently not. But that
          goes to show you what the
          people who have had experience
          with this think of what we're
          doing in British Columbia.

INT. ALEXANDRA MORTON'S LAB — NIGHT

Alexandra is working on her computer.

                    ALEXANDRA MORTON (O.S.)
          I got on the Internet and found
          some . . .

WIDER — She's typing.

                    ALEXANDRA MORTON (O.S.)
          . . . Norwegian scientists that
          had been studying sea lice for
          a long time,

OVER HER SHOULDER — We see her screen, filled with emails.

                    ALEXANDRA MORTON (O.S.)
          . . . and I sent a little timid
          email off to a bunch of them,
          saying,

RESUME — Morton interview.

A chum salmon juvenile (*Oncorhynchus keta*) was sampled on June 16, 2008, near a fish farm in the Broughton Archipelago. Most of the sea lice on this fish are salmon lice (*Lepeophtheirus salmonis*) with a few common sea lice (*Caligus clemensi*). Large infestations of sea lice on juvenile salmon have become common in areas with open net pen fish farms. (Photo credit: Alexandra Morton)

A juvenile pink salmon captured on July 21, 2014, was found to be infected with gravid salmon lice. A gravid female salmon louse has long egg strings that contain hundreds of eggs. The salmon louse's life cycle ranges from 8 to 9 weeks at 6°C and as little as 4 weeks at 18°C. So, in one calendar year, sea lice can produce between 6 and 12 generations of offspring when they have suitable hosts. (Photo credit: Alexandra Morton)

Sea lice graze on the mucus, skin and blood of their host. And when the host is a juvenile fish, the skin becomes compromised and the fish becomes susceptible to disease. This fish was collected near Quadra Island on June 5, 2005. (Photo credit: Alexandra Morton)

Alexandra Morton counts sea lice on juvenile fish sampled during one of her studies of sea lice loads on juvenile salmon. Her lab and home in Echo Bay, BC, in the Broughton Archipelago are now home to the Salmon Coast Field Station, a scientific charity continuing research into wild fish health. The facility continues to operate today. (Photo credit: Helen Slinger)

                    ALEXANDRA MORTON
               . . . "I've got juvenile salmon
               covered in sea lice, ah, could
               you tell me how to study them?

INT. ECHO BAY RESEARCH STATION - DAY

Alexandra is now at her computer, typing an email.

                    ALEXANDRA MORTON (O.S.)
               How do I figure out where these
               lice are from?"

THROUGH THE WINDOW - Alexandra is looking at a photo on her computer.

                    ALEXANDRA MORTON (O.S.)
               Only one answered me. And he
               goes, "Do you have fish farms
               in the area?" So I said, "Yes."
               And he emails me back . . .

RESUME - Morton interview.

                    ALEXANDRA MORTON
               . . . and he goes, "My
               suggestion to you is you
               drop this." He said, "Your
               government's not going to like
               it and the industry's not going
               to like it. This is a very big
               thing."

INT. ECHO BAY HOME — NIGHT

Alexandra is at her computer, reading the email.

          ALEXANDRA MORTON (O.S.)
    "You're not going to want to
    deal with this."

                NARRATOR
    Morton wasn't deterred by the
    Norwegian scientist's warning.
    After all, this is Canada.

INT. ECHO BAY (10 YEARS AGO) — DAY

Alexandra is in the research centre and is on the phone.

                NARRATOR
    How bad or political could
    it be?

          ALEXANDRA MORTON (O.S.)
    I phoned up DFO and I said,

RESUME — Morton interview.

            ALEXANDRA MORTON
    "You know, we've got a big sea
    lice problem here." And they
    said, "Oh, could you send us
    some?" So, I'm like okay, so I
    got them some, and then they
                    (MORE)

ALEXANDRA MORTON (CONT'D)
were like, "Busted. You're fishing without a licence."
(She laughs.)
And they sent an officer to my door and said ah, "Ah, you know, you, you might go to jail for this." And I'm like, "What? What? You guys told me that . . . You didn't tell me." . . . I . . . I shoulda thought. Of course you need a fishing licence for juvenile fish. I had a fishing licence for older salmon.

INT. DANIEL PAUL'S OFFICE — DAY

Establish. Interview. Dr. Daniel Pauly is a fisheries scientist and professor at the University of British Columbia. He leans back in his chair.

DANIEL PAULY
So basically she as a citizen, she reported about sea lice and it was expected the cop — in this case the DFO — to come and, and do something about it. And instead she got in trouble. She, people said, you didn't see what you saw.

INT. ECHO BAY RESEARCH LAB (1980s) — DAY

Alexandra is looking at a salmon smolt under a dissecting microscope.

> ALEXANDRA MORTON (O.S.)
> They were never actually trying to be helpful.

RESUME — Morton interview.

> ALEXANDRA MORTON
> Were they angry that I was looking, or were they angry that I was reporting what I saw? I wasn't too sure, you know, which of those two things was really my crime.

EXT. ALEXANDRA MORTON'S LAB — NIGHT

Over her shoulder. Wearing reading glasses, she flips through a binder full of documents.

MONTAGE — A sample of the pages Alexandra wrote and received.

> NARRATOR
> Morton wrote many letters to provincial and federal politicians and senior bureaucrats, telling them that sea lice from fish farms . . .

HANDHELD — Alexandra riffles through the pages in a binder.

> NARRATOR
> . . . were causing a serious problem for wild fish.

> ALEXANDRA MORTON (O.S.)
> When all of this . . .

RESUME — Morton interview. We can see that the wind is blowing outside the window behind her during the interview.

> ALEXANDRA MORTON
> . . . you know, slowly dawned on me, everything I was saying . . .

LETTER — Brian Tobin to Alexandra Morton, July 26, 1994.

THE WORDS — "Dear Ms. Morton" light up, followed by "no firm evidence that aquaculture."

> ALEXANDRA MORTON (O.S.)
> . . . to DFO about the algae blooms, the displacement of whales, which is a direct violation of the *Fisheries Act*, ah . . .

NEXT LETTER — Brian Tobin to Alexandra Morton, November 9, 1994.

THE WORDS — "Dear Ms. Morton" light up, followed by "concluded that aquaculture does not pose a threat."

> ALEXANDRA MORTON (O.S.)
> . . . the disease transmission to these coho, and other things,

NEXT LETTER — David Anderson to Alexandra Morton, February 23, 1998.

THE WORDS — "there is no evidence to suggest that fish farming adds significantly" leap off the page and light up.

> ALEXANDRA MORTON (O.S.)
> . . . they kept saying to me, "Dear Ms. Morton, there is no evidence of whale displacement,

NEXT LETTER — David Anderson to Alexandra Morton, September 4, 1998.

THE WORDS — "Absolutely no evidence for the involvement of fish diseases from aquaculture" leap off the page and light up.

> ALEXANDRA MORTON (O.S.)
> . . . of transfer of disease," of you know, whatever the concern was. And I began to recognize that "no evidence" . . .

NEXT LETTER — Subject — Fish Health Issues. January 18, 1999.

THE WORDS — "there is no evidence" leap off the page and light up.

>   ALEXANDRA MORTON (O.S.)
>   . . . was appearing in all of
>   these letters. I was like, "Oh,
>   they want evidence."

EXT. BROUGHTON ARCHIPELAGO — DAY

Alexandra is in her boat, collecting samples.

>   NARRATOR
>   Morton began to gather evidence
>   by studying juvenile pink and
>   chum salmon migrating past her
>   home at Echo Bay.

EXT. ANOTHER BEACH — DAY

Alexandra is working with a helper. They are pulling in a beach seine.

>   ALEXANDRA MORTON (O.S.)
>   We have to take this seriously.
>   This is . . . this is no longer
>   anecdotal. This is scientific
>   evidence.

RESUME — Morton interview.

Minister of
Fisheries and Oceans

Ministre des
Pêches et des Océans

JUL 2 6 1994

Ms. Alexandra Morton
Raincoast Research
Somoon Sound, British Columbia
BC V0P 1S0

Dear Ms. Morton:

Thank you for your recent letters in which you raise a number of issues related to the British Columbia salmon aquaculture industry.

My staff have reviewed these letters carefully and note that while you raise many points and have provided material from Ireland and elsewhere, you provide no firm evidence that aquaculture in British Columbia is the problem that you perceive it to be.

Your observations on the abnormal wild chinook is useful and I am pleased to hear that you were able to send samples to the Pacific Biological Station for analysis.

I regard aquaculture as a new but important component of the coastal and inland waters of Canada, an industry that provides needed income for coastal communities. It is apparent that you do not view aquaculture in this light and I believe that we will have to agree to disagree on the subject. However, I appreciate your concern for the environment of the Broughton Inlet.

Sincerely,

Brian Tobin

BT/tc

cc   Mary Russell

Ottawa, Canada K1A 0E6

Brian Tobin, the Minister of Fisheries and Oceans, responds to several letters written to the department by Alexandra Morton. Both the provincial and federal governments said that fish farms were not impacting wild salmon. (Credit: Government of Canada)

November 11, 1994

Minister Tobin
Minister of Fisheries
House of Commons
Ottawa, ONT
K1A 0E6

Dear Minister Tobin:

Please excuse this letter arriving so soon after the last one, but several more papers have arrived of which I thought you should be made aware. They further address the impact of fish farming ecologically and economically. The following information is contained within them.

> Tuomi believes that the development of the salmon farming industry would end the traditional commercial fishery for salmon in Canada. Tuomi. 1991.

> Even though there have only been a few investigations on the subject, the results indicate that fish farming affects natural fish communities. Horstad & Christensen 1991.

> Licenses will not be issued for farming salmon at localities which are connected with or near important salmon rivers. The reasons for this are the risk of disease infection both from farmed to wild and visa versa, and that potential escapees, when mature, tend to return to the area from which they escape. Bergan et al. 1991.

> The importation of exotic species or disease organisms poses the greatest environmental risk of mariculture since the consequences may be widespread and irreversible. Windsor & Hutchinson. 1990.

I will continue to send you information in the hopes that you will review your position that "the industry (salmon farming) does not pose a threat to wild salmon stocks".

Sincerely,

Alexandra Morton
cc: Moe Sihota
    U.F.A.W.U.
    West Coast Environmental Law Assoc.
    E. Warnock
    B. Ludwig
    H. Vogt

Alexandra Morton wrote many letters to various officials and politicians at both the federal and provincial levels of government. She believed that these letters would prompt a thorough investigation of the problems she was seeing in the environment. But the response from government at the time was to deny publicly that there were any impacts. (Credit: Province of BC)

                    ALEXANDRA MORTON
               But that didn't work either.
                    (She giggles.)
               Then they began to say, you
               know, crazy things like, ah,

INT. ECHO BAY LAB — DAY

On Alexandra's hands. She tugs at a fry with
lice on it.

                    ALEXANDRA MORTON (O.S.)
               . . . like I was sticking lice
               on fish, and, ah, because . . .

PULL BACK — Alexandra is examining a fry
with sea lice through a microscope and then
through a loop.

                    ALEXANDRA MORTON (O.S.)
               . . . my funding at some point,
               some of it came from the United
               States, I must be working for
               the Alaskan industry. They
               said, their results were
               different. After nine years of
               that,

RESUME — Morton interview.

                    ALEXANDRA MORTON
               . . . um, I turned my house
               into a research station because
               I could see that it wasn't
                         (MORE)

ALEXANDRA MORTON (CONT'D)
going to be enough for just
me to be doing this science. I
needed other scientists there.

EXT. ECHO BAY RESEARCH LAB — DAY

THROUGH THE WINDOW — Alexandra is working at her computer.

RESUME — Morton interview.

ALEXANDRA MORTON
I started with Dr. John Volpe
at the University of Victoria.
Worked on "What Happens to All
These Atlantics That Escape?"

HEADLINE — The *Fisherman*, "Gillnetters Shocked at Haul of Farm Fish," October 24, 1988.

RESUME — Pauly interview.

DANIEL PAULY
Well, first of all, I wouldn't
call the escape "escapes." I
would call them release. Why do
they always escape? It turned
out that they escape very
frequently because the net that
you put around your farm has
big meshes and they can, the
(MORE)

                    DANIEL PAULY (CONT'D)
               runts, the ones that have not
               grown in the previous phase,
               ah, they can escape.

HEADLINE — The *Fisherman*, "Salmon Farm
Escapes Demand DFO Action," February 17,
1989.

                    DANIEL PAULY (O.S.)
               When I came to Canada in '94,
               I remember the newspapers were
               saying . . .

HEADLINE — The *Fisherman*, "Escape of
Atlantics the 'Largest Ever' in B.C. Fish
Farms," August 23, 1993.

                    DANIEL PAULY (O.S.)
               . . . it is impossible that any
               fish escapes.

INT. STREAM — DAY

Atlantic salmon hover in the gentle current.

                    DANIEL PAULY
               And then fish escaped. And then
               they would not survive.

RESUME — Pauly interview.

Province of British Columbia
Ministry of Environment
FISHERIES BRANCH

Parliament Buildings
Victoria
British Columbia
V8V 1X5

TO: P Chamut

002414 NOV 14 10:35   File: 0215

November 7, 1986

Mr. Bruce A. Hackett
Assistant Deputy Minister
Financial Assistance Programs
Ministry of Agriculture
and Fisheries
4th Floor, 808 Douglas Street
Victoria, British Columbia
V8W 2Z7

Dear Mr. Hackett:

Re: Federal-Provincial Draft Policy on Salmonid Imports

Thank you for your October 17 comments on the draft policy, which I understand includes both Agriculture and Fisheries, as well as industry concerns. I also have independent comments from other elements of the industry as well as commercial and sport fishing groups.

I trust that you and your staff fully appreciate that this question of the introduction of exotic races of salmonids into British Columbia is probably the most critical issue ever to face the maintenance of wild salmonid stocks, as well as the over 400 million juvenile trout and salmon produced by government sponsored enhancement each year. These fisheries had a collective value in 1985 approaching one billion dollars comprised of $512 million landed value of the commercial salmon findings, $77 million economic value (1985 non-resident expenditures and resident willingness to pay) of the saltwater salmon sport fishery, and $88 million economic value of the freshwater trout and salmon sport fishery. These values represent 104,000 metric tons of salmon in the commercial fisheries, 2.8 million taken in the tidal sport fisheries, and 8 million trout and kokanee taken in the non-tidal sport fisheries. In this light, I think you can understand our extreme caution with any endeavour that puts to risk such a valuable resource.

Fundamental to our import policy is our deep concern that we might receive diseases or strains of diseases that are not now present in British Columbia and that will cause extreme problems in government and private culture and may impact on wild production as well. These include IPN, VHS and Whirling Disease from Europe and Type II IHN and strains of vibrio and furunculosis from the Great Lakes and the Pacific Northwest.

...2

000082

Dave Narver, director of the BC Ministry of Environment, Fisheries Branch, and other senior officials at both levels of government were initially opposed to and expressed many concerns about allowing aquaculture operations that involved importing live fish or eggs from other parts of the world. (Credit: Province of BC)

Minister of
Fisheries and Oceans

Ministre des
Pêches et des Océans

Ottawa, Canada K1A 0E6

SEP 4 1998

Ms. Alexandra Morton
Raincoast Research
Simoom Sound, British Columbia
V0P 1S0

Dear Ms. Morton:

Thank you for your letter of May 30, 1998, concerning the risks of Infectious Salmon Anemia (ISA) spreading to wild Pacific salmon stocks in British Columbia. Specifically, you recommended a ban on the importation of Atlantic salmon for salmon farming in British Columbia.

As indicated in your letter, there is a growing body of information about ISA based on studies in Norway, and more recently in New Brunswick. Some of the relevant facts are summarized below:

- Atlantic salmon (*Salmo salar*) is the only species known to be susceptible to the disease, but brown trout (*Salmo trutta*) and rainbow trout (*Oncorhynchus mykiss*) have been shown experimentally to be carriers of the virus.

- Data to date suggest that the infection is only seen in fish held in seawater (e.g. in marine salmon farms) or in fish exposed to seawater. ISA has never been found in freshwater hatcheries.

- ISA is not known to occur in wild fish, either in Canada or Norway (where the disease was first detected).

- ISA has been reported very recently in Scotland for the first time, again only in farmed fish in coastal waters. The disease has been confirmed at two sites and is suspected at another 7 sites (not 72 farms as stated in your letter).

.../2

Canada

In spite of growing evidence to the contrary, politicians maintained that there was a lack of evidence that fish farms were having a serious impact on wild salmon populations. (Credit: Government of Canada)

                    DANIEL PAULY
          Why not? They're fish.
          Otherwise they are in the
          water. Why, why would they not
          survive? And then, they would
          not find a river to go. And
          then they were found in rivers.
          And they would never spawn. And
          then, uh, spawners were found
          by John Volpe.

INT. STREAM — DAY

Salmon spawning in the river.

                    JOHN VOLPE (O.S.)
          When you bring these farms in
          tight you've got chronic and
          constant release of fish in a
          particular area. The windows of
          opportunity for invasion will
          shrink and expand through time.

RESUME — Volpe interview.

                    JOHN VOLPE
          Some years, you know, conditions
          aren't very good for Atlantic
          salmon. In other years the
          conditions are, are, are better.
          If you've got a lot of farms
          in a small area and you're
          constantly putting out those
          fish, that ensures that those
                         (MORE)

>   JOHN VOLPE (CONT'D)
>   fish are in the area when that
>   window expands to its maximum
>   and . . .

INT. STREAM — DAY

An Atlantic salmon digs a redd in the gravel.

>   JOHN VOLPE (O.S.)
>   . . . then they can take
>   advantage of it.

>   BRAD HOPE (O.S.)
>   Trying to keep track of what
>   you had,

RESUME — Brad Hope interview.

>   BRAD HOPE
>   . . . you'd think you knew the
>   numbers, or you'd count every
>   mortality, count the smolts
>   that went in rather,

INT. FISH FARM SHACK — DAY

Two men look at a chart.

EXT. FISH FARM PEN — DAY

Another farmer looks into a shaded rearing pen.

CLOSER — Smolts swim inside the holding pen.

> BRAD HOPE (O.S.)
> . . . they should add. And they never did. They were always many, many less than you'd think and so we started off in areas where there was heavy tides, which was great in some ways,

A BIT WIDER — The current whips the smolts around in their pen.

> . . . it was moving oxygen through, was keeping everything clean below the nets, so we could keep things nice and clean but it certainly promoted the growth of mussels . . .

RESUME — Brad Hope interview.

> BRAD HOPE (O.S.)
> . . . and, and what not, but also brought logs and chunks of things that would get caught in the net and tear a net, and you'd find a hole in the net and you had no idea of knowing whether you'd lost half your stock or a third of it or none of it.

HEADLINE — The *Fisherman*, "350,000 Farm Salmon Escape as Winter Storm Smashes Pens," February 17, 1989.

NARRATOR
There have not been just a
few farm fish escapes, as some
politicians have claimed. There
have been hundreds of thousands
over the years, and Atlantic
salmon have been turning up in
streams up and down the Pacific
coast.

RAFE MAIR (O.S.)
If you go right back to the
beginning of this . . . this
issue,

RESUME — Mair interview.

RAFE MAIR
. . . ah, the fish weren't
going to escape from the
nets. If they did escape, they
weren't going to be able to
survive in the wild. If they
did survive in the wild, they
would never get into the
rivers. If they did get into
the rivers, they would never
spawn. If they got in the
rivers and spawned, said the
Department of Fisheries and
Oceans, we'll get rid of them
once they're in the rivers, to
which my reply is, yeah, we can
bank on that. You've been doing
(MORE)

>   RAFE MAIR (CONT'D)
>   that for centuries. That's
>   something you're good at.

HEADLINE — The *Fisherman*, "Puget Seiners Land Atlantics," October 24, 1988.

INT. CREEK — DAY

Atlantic salmon pass through frame.

>   NARRATOR
>   There are many examples of
>   species being introduced
>   from one area of the world
>   to another with disastrous
>   results.

INT. PACIFIC COAST ESTUARY (AERIAL) — DAY

A number of fish farms are distributed around the bays and inlets along the coast.

EXT. PACIFIC COAST FISH FARM — DAY

An aerial view.

>   NARRATOR
>   In 1981, scientists at the
>   Department of Fisheries and
>   Oceans were so concerned about
>   the introduction of Pacific
>   salmon . . .

REPORT COVER — "Behavioural Interactions Between Coho Salmon (Oncorhynchus kisutch), Atlantic Salmon (Salmo salar), Brook Trout (Salvelinus fontinalis), and Steelhead Trout (Salmo gairdneri) at the Juvenile Fluviatile stages." September, 1981.

                NARRATOR
. . . to fish farms on the east coast, they examined the competitive fitness of several salmonid species. They found that Pacific steelhead were the only species more aggressive . . .

INT. PHOTARIUM — DAY

A juvenile steelhead is resting on the bottom of the tank.

                NARRATOR
. . . than Atlantic salmon.

INT. CREEK — DAY

One Atlantic salmon chases another in the frame.

              JOHN VOLPE (O.S.)
When we find an adult Atlantic salmon in the river, sometimes the, the snouts are scuffed up or the fins are worn,
                  (MORE)

                    JOHN VOLPE (CONT'D)
          suggesting that, that fish is
          a recent escapee, but many
          are, you know, look very, very
          healthy, and very,

RESUME — Volpe interview.

                    JOHN VOLPE
          . . . in good shape. And you
          wonder, is that this is a fish
          that escaped a while ago and,
          and all the signs of being in
          captivity have healed over,
          or, you know, is this a fish
          actually returning to its natal
          river as a, as a truly wild
          fish?

EXT. ECHO BAY — DAY

The shoreline is spotted with a few modest homes and wharves.

EXT. ROCKY SHORE — DAY

Two Raincoast researchers, Scott Rogers and Kristen Frake, are setting up their boat and a beach seine to sample fry.

FROM THE BOAT — Kristen holds one end of the net while Scott releases the net into the water.

WIDER — The net is now strung across a small bay.

EXT. SHORELINE — LATER

The net is now a tight circle and the two women pull the ends up onto shore.

>   NARRATOR
>   Alexandra's Echo Bay home . . .

ON KRISTEN — She pulls hard on the net.

>   NARRATOR
>   . . . was converted into a
>   research station to study the
>   sea lice issue . . .

OVER THEIR SHOULDERS — They use a small dip net to transfer salmon fry from the beach seine into a sampling pail.

>   NARRATOR
>   . . . and a new group of
>   scientists began to ask tough
>   questions . . .

THE WOMEN — sit side by side, counting and recording the numbers of sea lice on their sample.

>   NARRATOR
>   . . . about the impact sea lice
>   were having . . .

ON SCOTT — She captures a small fry and places it in a Ziploc bag to count the sea lice on it.

> NARRATOR
> . . . on wild salmon.

TIGHT — A large sea louse is on a fry.

> MARTY KRKOSEK (O.S)
> The first one was, what's the role of the salmon farms in the sea lice infestations that . . .

RESUME — Krkosek interview.

> MARTY KRKOSEK
> . . . have been happening here? The second one was, what's the effect of sea lice on the survival of juvenile salmon? And then, the third and most important question was, if this is happening year after year, what are the implications for salmon populations?

BACK TO SCOTT AND KRISTEN — They continue to sample and count sea lice on fry.

> NARRATOR
> The first impact researchers noticed was the physical damage sea lice have on small fish.

RESUME — Orr interview.

                    CRAIG ORR
          If you have a high intensity
          of infections, several lice per
          fish, it can cause all kinds
          of problems with . . . with
          juvenile fish. It can . . .

EXT. OCEAN — DAY

Looking down into the water next to the fish
farm. Fry are flashing as they swim.

                    CRAIG ORR (O.S.)
          . . . cause them to do this
          flashing behaviour which we've
          seen in the, the Broughton
          Archipelago where they're,
          they're, they're covered with
          lice and they're flashing . . .

ANOTHER ANGLE — of the same farm.

                    CRAIG ORR (O.S.)
          . . . and, ah, any fish that
          starts flashing like that, you
          know,

INT. OCEAN — DAY

Small chum and pink fry swim along a wall.

                    CRAIG ORR (O.S.)
          . . . becomes a beacon to
          predators, like coho salmon and
          cutthroat and birds and things
          like that. It impairs their
          swimming ability. That's been
          shown in several laboratory
          studies.

PHOTO — A small pink fry in a photarium. It is covered in sea lice.

                    CRAIG ORR (O.S.)
          But the biggest problem is at
          high levels of infestation they
          can cause mortality . . .

ON A DEAD FRY — The life has been sucked right out of it.

                    CRAIG ORR (O.S.)
          . . . by sucking most of the
          fluids out of these fish.

                    JOHN VOLPE (O.S.)
          When you come out of the
          river . . .

RESUME — Volpe interview.

Very heavy sea lice loads have been discovered on juvenile salmon since the mid-1980s. These pink salmon are mostly infected with *Caligus clemensi*, the common sea louse. (Photo credit: Alexandra Morton)

Two sockeye juveniles (*Oncorhynchus nerka*) are infected with sea lice. The top fish has two large salmon lice (*Lepeophtheirus salmonis*) with two common sea lice (*Caligus clemensi*) on its upper back. The bottom fish has two common sea lice on its side. Scientists have shown that juvenile salmon have not yet developed thick scales and are more susceptible to skin damage from sea lice. (Photo credit: Alexandra Morton)

                    JOHN VOLPE
          . . . and you're, you're
          that . . .
          (demonstrates with his fingers)
          . . . big, right, which is the
          size of a pink or a chum, chum
          salmon, right?

INT. STREAM — DAY

Small chum fry come warily into frame.

                    JOHN VOLPE (O.S.)
          You have no scales even, you
          know, no physical defences
          against these things . . .

PHOTO — A pink or chum fry.

INT. PHOTARIUM — DAY

Pink salmon covered in lice.

                    JOHN VOLPE (O.S.)
          . . . and the size of the louse
          when it matures on that fish is
          equivalent to you of having a
          40 pound parasite on your back
          chewing through your skin.

MONTAGE — Pink and chum fry at their earliest stages.

EXT. STREAM — DAY

Chum and pink fry mingle together in a small school.

SHOT — Three small chum fry swim frantically in the current.

> MARTY KRKOSEK (O.S.)
> So for the smallest fry with an adult stage louse on it, I'd say one is lethal.

MONTAGE — A mature sea-run cutthroat trout with sea lice on it in a photarium.

> MARTY KRKOSEK
> As the fish grow and as the lice get smaller, it gets a little bit more complicated. It would take more lice to kill an individual fish.

EXT. OCEAN — DAY

No fish farms in sight.

RESUME — Krkosek interview.

> MARTY KRKOSEK (O.S.)
> When there's no salmon farms there, the adults, they go up the rivers, they spawn . . .

EXT. ADAM'S RIVER SPAWNING CHANNEL — DAY

A male sockeye salmon is ready to spawn.

EXT. ADAM'S RIVER BANK — LATER

There are dead sockeye salmon stacked along the shore, spawned out.

> MARTY KRKOSEK (O.S.)
> . . . and they die, and it's about, you know, three, four, five months later, maybe six months later,

INT. WEAVER CREEK — DAY

Chum and sockeye salmon fry dart around, fresh out of the gravel.

> MARTY KRKOSEK (O.S.)
> . . . before the juveniles come out of the rivers and enter the ocean.

INT. OCEAN — DAY

No fish, but a clean environment.

> MARTY KRKOSEK (O.S.)
> So there's several months there where the ocean cleans up of whatever pathogens would be associated with the salmon.

RESUME — Krkosek interview.

                    MARTY KRKOSEK
          There isn't any host there,
          naturally.

INT. RIVER — DAY

Eulachon school in a pool.

                    NARRATOR
          Many marine fish species spawn
          in or near fresh water. And
          not all of the fish die after
          spawning. But the sea lice do
          because fresh water is lethal
          to them.

A SALMON FRY — It's swimming in very shallow water.

PHOTO — Two pink salmon juveniles with sea lice on them.

PHOTO — A single pink salmon smolt with at least 15 sea lice on it.

                    NARRATOR
          This serves to keep sea lice
          populations in check in the
          wild.

INT. PACIFIC BIOLOGICAL STATION NET PEN (1970s) — DAY

A large chum swims toward camera. The image freezes for a moment and an arrow points at the sea louse. The fish moves forward again and there are two more sea lice on the back of the fish.

>   NARRATOR
>   But by keeping salmon in
>   pens year round, sea lice
>   populations multiply quickly,
>   upsetting nature's balance.

INT. PHOTARIUM — DAY

A sockeye smolt has two gravid female salmon lice on its side.

>   JOHN VOLPE (O.S.)
>   A female sea louse can produce,
>   you know, say, 400 eggs, right,
>   you've got a million fish. You
>   know, each of those fish might
>   have a couple of female louse
>   on them, you know, the number
>   of,

INSERT — A jar of sea lice juveniles from plankton tow.

>   JOHN VOLPE (O.S)
>   . . . of larvae that are being
>   emitted from that farm now
>   (MORE)

                    JOHN VOLPE (O.S.) (CONT'D)
               becomes tremendous and it's
               through that cloud of effective
               larvae that these young fish
               have to swim.

EXT. FISH FARM — DAY

Water laps against the pens.

RESUME — Orr interview.

                    CRAIG ORR
               In fact, we've been out on
               salmon farms in the Broughton
               Archipelago counting lice
               because we didn't believe the
               numbers.

PHOTO — Inside a sampling net of a young
chinook with sea lice on it.

                    CRAIG ORR (O.S.)
               And, ah, we wanted some
               transparency on these numbers.

PHOTO — Several mature salmon sea lice on
sockeye smolts inside the photarium.

                    CRAIG ORR (O.S.)
               And we may have only found
               three or four lice per fish,

PHOTO — A young chinook with a bunch of
herring.

This Pacific herring (*Clupea pallasii*) is infected with the common sea louse (*Caligus clemensi*). Pacific herring are not usually susceptible to salmon lice, but the common sea louse can infect many species of fish. Sea lice infestations on juvenile herring were unreported prior to salmon farming on the BC coast. (Photo credit: Alexandra Morton)

Many species of fish often congregate together. So when there are high levels of sea lice in the environment, they can easily jump between species. (Photo credit: Jody Eriksson)

                    CRAIG ORR (O.S.)
               . . . but when you have 750,000
               fish and you have these sea
               lice, which are very short life
               cycle, in fact . . .

PHOTO — Four herring with *Caligus clemensi*
sea lice on them.

                    CRAIG ORR (O.S.)
               . . . it may be 35 to 40 days
               between egg and adult in their
               life cycle.

PHOTO — A mature sea-run cutthroat trout with
sea lice near its tail.

                    CRAIG ORR (O.S.)
               You can be producing millions
               or billions of lice . . .

PHOTO — A pink salmon juvenile, covered from
one end to the other with gravid salmon sea
lice, in a photarium.

                    CRAIG ORR (O.S.)
               . . . per farm over the course
               of a year if untreated.

PHOTO — A pink salmon juvenile with at least
13 sea lice on one side of its body.

INT. ESTUARY — DAY

A large school of juvenile chum fry dart through the shallow water.

> MARTY KRKOSEK (O.S.)
> The conclusions were pretty dramatic. What we found is that in the absence of salmon farms, you'd typically see only about five percent of the juvenile salmon infected with lice.

Moments later they dart back in the opposite direction.

INT. BUCKET — DAY

A sockeye smolt swims with a louse on its back.

> MARTY KRKOSEK (O.S.)
> So only five out of 100 fish would have one louse. And what was happening here . . .

EXT. WEST COAST (AERIAL) — DAY

A ten-pen farm sits off a treed shoreline.

> MARTY KRKOSEK (O.S.)
> . . . during those years is we had, ah, you know, 75, 85, even over 90 percent infection rates . . .

EXT. FISH FARM (AERIAL) — DAY

A large farm with at least 18 pens runs parallel to the shoreline.

> MARTY KRKOSEK (O.S.)
> . . . after the, the fish have passed the salmon farms, and ah, you know, 20, 30, even up to 80 lice per fish.

ANIMATED MAP — Juvenile salmon migration.

The farm sites are marked with white dots. Salmon migration routes are illustrated, and as the fish pass the farms, the runs go from normal green, to reduced yellow, to critically red.

> MARTY KRKOSEK (O.S.)
> And these are lethal loads. And it's obvious that the lice were having a major impact on the survival of these fish.

RESUME — Krkosek interview.

> MARTY KRKOSEK
> You didn't really need sophisticated math to figure it out.

INT. HELICOPTER — DAY

The helicopter is flying over the Wakeman River.

                    CRAIG
          Now, here's where you will
          start seeing a few. See the
          action in the water down there?

                    TWYLA ROSCOVICH
          Ahhhh.

                    HELICOPTER PILOT
          The wrinkles.

                    TWYLA ROSCOVICH
          Yeah. I see the wrinkles. I
          don't see any fish, though.

                    CRAIG
          Exactly.

                    PILOT
          It used to be full of pinks
          through here. Remember this,
          Craig?

                    CRAIG
          Oh yeah.

                    PILOT
          It was nuts through here.
          Remember this, Craig?

CRAIG
Oh yeah.

PILOT
It was nuts through here . . .

CRAIG
Thousands, thousands of pinks down here.

ALEXANDRA MORTON — She's sitting quietly, looking down. She looks nervous.

TWYLA ROSCOVICH
Oh, this looks like a pretty good little spawning ground here, eh? If there was pinks?

PILOT
Yeah, this is the chinook spawning grounds right below you. If there was pinks, they'd be jumping all along here. The pinks used to sit right in here, and this is the chinook hole. You'll get 50 chinook hang off that rock there. We tried today, Craig. Zip. Nothing.

CRAIG
Yeah. Nothing.

PILOT
We tried the other day too. Nothin'.

                    MARTY KRKOSEK (O.S.)
               The results were . . .

RESUME — Krkosek interview.

                    MARTY KRKOSEK
               . . . arresting, you know. The,
               the change in the productivity
               here went from one that used
               to support commercial fisheries
               to one where wild salmon
               populations were simply no
               longer viable. They were on
               a rapid trajectory to local
               extinction, and ah, this was
               happening on the order of only
               about less than ten years, less
               than ten years from historical
               abundance to, ah, no more
               salmon.

                                        FADE TO:

EXT. WHARF — EVENING

Alexandra looks out into the channel near her home. Her dog is sitting at her feet.

                    ALEXANDRA MORTON (O.S.)
               It was up to somebody to go to
               court about this.

CLOSER — The wind blows her hair. She's wrapped up against the cold.

>   ALEXANDRA MORTON (O.S.)
>   You know, you make yourself
>   a target for legal action
>   yourself.

Alexandra exits and heads back to her home.

>   GREGORY MCDADE (O.S.)
>   Alex actually came to us first
>   looking to go directly against
>   the fish farms in terms of
>   a nuisance action for the
>   damage . . .

RESUME — McDade interview.

>   GREGORY MCDADE
>   . . . that they're doing to
>   commercial fishermen from
>   salmon. But as we got deeper
>   into looking at that question,
>   it became clear that there
>   was a much more fundamental
>   problem.

RESUME — Morton interview.

>   ALEXANDRA MORTON
>   He got back to me and he goes,
>   "You know, I don't think this
>   whole fish farming thing is
>   legal in Canada." I was, I
>   didn't understand him at
>   all . . . I was like, "It must
>   (MORE)

                    ALEXANDRA MORTON (CONT'D)
          be. You know, there's farms
          everywhere. It's been going
          on for 20 years. What do you
          mean it's not legal?" "Well,"
          he said, "It's not that it's
          illegal, it's just outside the
          law."

EXT. FISH FARM SITE — DAY

Alexandra and friends are in a boat.

ANOTHER BOAT — approaches Alexandra's boat.
It's a fish farm worker.

                    MAN
          Ah yeah. I'm going to have to
          ask, ask you to leave this
          leased property.

                    ALEXANDRA MORTON
          Well, that's the lease right
          there. Not here.

                    MAN
          Yeah, no, actually it's. . .
                    (wind)

He points to two places.

                    MAN
          . . . from that point over
          there on that line past that
          rock.

ALEXANDRA MORTON
You only have a licence of occupation. You're not allowed to really displace boats. Is there a reason we have to go?

MAN
No, but, it's just that uh . . .
(more wind)

EXT. LARGE FISH FARM — DAY

Establish.

GREGORY MCDADE (O.S.)
But really the case came down to quite a very simple proposition: Was a fish farm a fishery,

TIGHTER — on the floats around the farm.

GREGORY MCDADE (O.S.)
. . . in which case it had to be regulated by the federal government, or was it a farm, in which case the province could regulate it through aquaculture? In other words, was it private property or was it part of the public resource?

ANOTHER SHOT — We see the fish farm pens up close.

>   GREGORY MCDADE (O.S.)
>   So now the big question is,
>   what happens next? Right?
>   Because the federal government
>   had gotten out of the business
>   of,

RESUME — McDade interview.

>   GREGORY MCDADE
>   . . . of protecting the fish.

EXT. BC SUPREME COURT BUILDING — DAY

SUPERSCRIPT: February 9, 2009.

Alexandra is participating in an impromptu protest.

>   ALEXANDRA MORTON (O.S.)
>   Working with the province on
>   this has been very difficult,
>   and there's been no progress.
>   Our wild salmon are in worse
>   condition than they've ever
>   been. But as a biologist I see
>   indicators that they could be
>   restored.

EXT. COURT BUILDING (ARCHIVAL) — DAY

Alexandra is facing the media with Gregory McDade at her side.

>   ALEXANDRA MORTON
>   And I'm really looking forward
>   to getting onto that step
>   because I have a strong vision
>   as to how we could have wild
>   salmon on this coast for the
>   benefit of everybody.

RESUME — McDade interview.

>   GREGORY MCDADE
>   Fish farm lawyers suggested
>   that these fish in a fish farm
>   were more like chickens . . .

INT. GOVERNMENT HATCHERY — DAY

A man is working with a tray.

>   GREGORY MCDADE (O.S.)
>   . . . than they were like fish
>   because you kept them all
>   together in one place.

TIGHT ON — alevins swimming in the tray.

>   GREGORY MCDADE (O.S.)
>   So once the judge decided that
>   this was a . . .

EXT. POND — DAY

A number of adult salmon swim around in a brood stock pen.

> GREGORY MCDADE
> . . . fishery, it was obvious that it was federal jurisdiction.

INT. VANCOUVER APARTMENT — DAY

Establish. Interview. John Fraser, retired politician and lawyer, was the Minister of Fisheries when the Atlantic salmon populations were crashing on the east coast of Canada.

TITLE: John Fraser, Former Minister, Fisheries and Oceans and Speaker of the House of Commons.

> JOHN FRASER
> It is my conviction that in almost every case, you have to sit down and say who shoulda been in charge? Who should've had a responsibility? And why didn't someone have the responsibility? And that's your answer.

FRONT PAGE — The Supreme Court of BC petition.

> NARRATOR
> Morton won her case and overturned the MOU signed in 1988. And it did change one
> (MORE)

                    NARRATOR (CONT'D)
          important thing. Someone is
          now in charge when it comes to
          aquaculture in Canada. But what
          would the federal government do
          with its reaffirmed power?

EXT. DFO PACIFIC HEADQUARTERS VANCOUVER — DAY

Establish.

ON A FENCE — A DFO sign says "No
Trespassing."

                    GREGORY MCDADE (O.S.)
          Having allowed this industry
          to get a toehold in our oceans,
          both levels of government are
          very reluctant . . .

RESUME — McDade interview.

                    GREGORY MCDADE
          . . . to just make it go away.

INT. AQUACULTURE CONFERENCE — DAY

A poster in the foreground is promoting
Canada as "A Leader in Sustainable
Aquaculture."

WIDER — Bureaucrats are working their booth.

LOOKING UP — at Canada's banner hanging from
the roof.

No. S083198
Vancouver Registry

IN THE SUPREME COURT OF BRITISH COLUMBIA

Between

ALEXANDRA B. MORTON,
PACIFIC COAST WILD SALMON SOCIETY,
WILDERNESS TOURISM ASSOCIATION,
SOUTHERN AREA (E) GILLNETTERS ASSOCIATION, and
FISHING VESSEL OWNERS' ASSOCIATION OF BRITISH COLUMBIA

Petitioners

And

MINISTER OF AGRICULTURE AND LANDS
The ATTORNEY GENERAL OF BRITISH COLUMBIA
on behalf of THE PROVINCE OF BRITISH COLUMBIA, and
MARINE HARVEST CANADA INC.

Respondents

ARGUMENT OF THE PETITIONERS

This is the cover for the Supreme Court of British Columbia court case known as the Morton decision. The case overturned the MOU between the Province of British Columbia and the Government of Canada that handed regulation of the aquaculture industry to the province. This ruling decided that it was illegal for the Government of Canada to abandon its responsibility. Since this decision was made, regulation of the industry has gone back to the Government of Canada and its Department of Fisheries and Oceans. (Credit: Province of British Columbia)

A MAN — demonstrates a fish processing conveyor belt.

> NARRATOR
> And wherever fish farm companies are in the world, they have been effective at persuading politicians to go soft on regulation while telling their opponents that . . .

CLOSE UP — on Canada's brochures.

A LARGE FISH FARM — fills a bay on the BC coast.

> NARRATOR
> . . . they are merely following the rules as set out by regulators.

PANNING — A fish farm pen with a tangle of security nets over it.

> NARRATOR
> And with all the lobbying pressure and the downsizing of government, some politicians thought it would be a good idea to let companies regulate themselves.

ANOTHER LARGE FARM — with a crane on a barge tethered to it as it unloads cargo.

                    DANIEL PAULY (O.S.)
               The reason why self-regulation
               of these . . . of these farms
               is not possible.

RESUME — Pauly interview.

                    DANIEL PAULY
               They must be regulated from
               outside. They must be, they
               must be forced to behave
               themselves. They're behaving
               like three-year-old, three-,
               four-year-olds. They have no
               morals.

EXT. CHILEAN LANDSCAPE — DAY

A flag flies in the foreground.

                    ALEJANDRO BUSCHMANN RUBIO (O.S.)
               During the last months here
               in Chile, there were several
               claims that the . . .

EXT. CHILEAN STREET — DAY

A faded fish farm sign fills the frame.

                    ALEJANDRO BUSCHMANN RUBIO (O.S.)
               . . . country or the government
               did not regulate enough.

EXT. CHILEAN FISH FARM — DAY

This farm is even larger than the farms in Canada.

EXT. CHILEAN FORESHORE — DAY

Field interview. Alejandro is standing on the shore near a marina.

TITLE: Dr. Alejandro Buschmann Rubio, Professor, Marine Biology, University of Los Lagos.

> ALEJANDRO BUSCHMANN RUBIO
> Ok, that's true. But it's also relevant to say that the companies did also make a very strong lobby against regulations. So, at the end, you have a weak government and on the other sides you have strong pressure from companies.

FADE TO:

EXT. DFO PACIFIC HEADQUARTERS VANCOUVER — DAY

Establish.

> GREGORY MCDADE (O.S.)
> The federal government seems to care more about the income from aquaculture and industry.

ON THE SIGN — Canada.

EXT. BC CORPORATE FARM — DAY

It's almost as big as the Chilean farm.

>   GREGORY MCDADE (O.S.)
> So when the industry says
> to the government, "We don't
> want you asking for disease
> information,"

INT. HARRISON RIVER — DAY

A silver, but dead, sockeye lies at the bottom of the river.

>   GREGORY MCDADE (O.S.)
> . . . that they've been very
> successful in convincing the
> federal government, so far, not
> to do that.

RESUME — McDade interview.

>   GREGORY MCDADE
> God knows somebody's got to
> start asking for information
> about disease.

                                        FADE OUT:

FADE IN:

MONTAGE — Aerial shots of BC's fish farms.

RESUME — Orr interview.

>   CRAIG ORR
> The sea lice don't just target
> pink salmon. They target
> any species of salmon they
> can find. Ah, fairly heavy
> infestations on, on juvenile
> chum. We've also seen them
> on . . . on coho. We've seen
> them on chinook. We've seen
> them on steelhead. Ah, we're
> also seeing them now on
> sockeye.

EXT. EAST COAST SALMON FARM — DAY

Establish. The large farm fills the foreground. The black nets of its design have an alien quality to them.

WIDER — Bird nets cover the circular pens. The farm is located in a channel or inlet.

>   NARRATOR
> The salmon louse also targets
> salmon in the Atlantic Ocean.
> But it gets even more complex,
> because there are types of sea
> lice that can target many, if
> not all, species of fish.

INT. FISH FARM — DAY

Dark images of salmon move through the murky water.

                    NARRATOR
          In the Atlantic Ocean,
          Caligus elongates is also a
          generalist and has been studied
          extensively and . . .

INT. ATLANTIC OCEAN — DAY

Cod swim around foraging.

                    NARRATOR
          . . . is now known to
          parasitize at least 80 species
          of fish.

PHOTO — A herring with two gravid *Caligus clemensi* sea lice on it.

                    NARRATOR
          It is also believed that the
          Pacific *Caligus clemensi* has a
          similar range of hosts and has
          been . . .

PHOTO — A pink salmon juvenile with a lot of *Caligus clemensi* sea lice.

PHOTO — Several herring in a photarium have *Caligus clemensi* sea lice on them.

PHOTO — A school of eulachon.

NARRATOR
. . . observed on salmon, herring, eulachon,

PHOTO — A ling cod with several *Caligus clemensi* sea lice.

NARRATOR
. . . ling cod and many other species.

PHOTO — Two *Caligus clemensi* sea lice are attached to a herring.

CRAIG ORR (O.S.)
Caligus is a species of concern in terms of,

RESUME — Orr interview.

CRAIG ORR
. . . of a louse and a parasite. Not so much, ah, because of its size. I mean, it's again,

PHOTO — A sockeye with *Lepeophtheirus salmonis*, salmon lice, on it.

CRAIG ORR (O.S.)
. . . it's much smaller than *Lepeophtheirus*, or *Leps* as we call them, and, and doesn't cause as much physical damage as on these fish.

PHOTO — A herring with three *Caligus clemensi* sea lice on it.

>			CRAIG ORR (O.S.)
>		But one of the things that
>		concerns us . . . is that . . .

PHOTO — Another herring with a *Caligus* on it.

>			CRAIG ORR (O.S.)
>		. . . *Caligus* are very mobile.
>		They can jump between fish
>		and . . .

PHOTO — Two herring, one with lice.

>			CRAIG ORR (O.S.)
>		. . . any animal that gets
>		around from one animal to
>		another can also transport
>		disease.

PHOTO — A single herring with two adult *Caligus* and at least two immature ones.

>			CRAIG ORR (O.S.)
>		It's called a vector of disease.
>		And we are very concerned
>		that *Caligus* is a vector for
>		diseases.

UNDER A MICROSCOPE — A close shot of the mouth parts of a sea louse.

                    NARRATOR
          As it turns out,

INT. NYLAND LAB — DAY

Dr. Nyland is reading documents on his computer.

                    NARRATOR
          . . . more than two decades ago,
          Norwegian scientists performed
          experiments . . .

INT. MICROSCOPE — DAY

CLOSE SHOTS OF SEA LICE.

                    NARRATOR
          . . . to find out if sea lice
          are disease vectors. They came
          to some startling conclusions.

FRONT PAGE — Diseases of Aquatic Organisms. "Mechanisms for transmission of infectious salmon anaemia (ISA)," A. Nylund et al. July 28, 1994.

                    NARRATOR
          Atlantic salmon smolts were
          exposed to the infectious
          salmon anemia virus by putting
          them in tanks with infected
          fish.

INT. RESEARCH TANK — DAY

Atlantic salmon adults swim around in the tank.

EXT. TANK — DAY

Looking down into the tank of Atlantic salmon.

> NARRATOR
> They found that uninfected smolts got the virus when the infected fish had sea lice on them,

INT. TANK — DAY

The fish swim along the wall of the tank.

> NARRATOR
> . . . but did not get the virus when sea lice were absent. This groundbreaking work should have set off the alarm bells for Canada's regulatory bodies. This study showed that . . .

INT. OCEAN — DAY

Sockeye smolts, some with sea lice, are caught in a beach seine for sampling.

                    NARRATOR
          . . . sea lice, by jumping
          from one fish to another, carry
          the disease and pass it on to
          the next host.

INT. RIVER — DAY

Mature chum salmon are nearly ready to spawn.

                    NARRATOR
          So this begs the question, if
          disease can travel through sea
          lice,

EXT. NORWEGIAN FISH FARM — DAY

Establish.

CLOSER ON THE NETS

                    NARRATOR
          . . . what is the potential
          reach of a disease outbreak at
          a fish farm?

DOCUMENT — Population genetic differentiation of sea lice (*Lepeophtheirus salmonis*) parasitic on Atlantic and Pacific salmonids: analysis of microsatellite DNA variation among wild and farmed host.

                    NARRATOR
          Genetic analysis of
          Lepeophtheirus salmonis
          discovered that this sea louse
          species is from one large
          population that extends right
          across the Atlantic Ocean.

ANIMATED MAP — The North Atlantic Ocean.

The range for *Caligus elongatus* fades up.

                    NARRATOR
          This means the salmon louse has
          a very large unimpeded range.
          And studies of the species-
          hopping *Caligus* have come to
          similar conclusions. So it's
          not a question of whether a
          sea lice disease will be spread
          throughout the range. It's more
          a question of how fast will it
          get there.

HEADLINE — The *Fisherman*, "Scottish Herring
Closure Legacy of Overfishing," September 8,
1978.

                    NARRATOR
          So when overfishing was blamed
          for the collapse of the herring
          and cod in Scotland, Norway
          and Canada, was it really
          overfishing?

DOCUMENT — Interactions of caligid ectoparasites and juvenile gadids on Georges Bank, Marine Ecology. September 1987.

> NARRATOR
> A study done by DFO scientists in 1987 may have discovered the beginning of a sea lice population explosion off the east coast of Canada.

PHOTO — An Atlantic salmon covered in *Caligus elongatus* sea lice.

PHOTO — A young cod hovering in the darkness.

> NARRATOR
> Sea lice numbers increased 360 percent on juvenile cod and 1400 percent on . . .

A JUVENILE HADDOCK — It swims slowly in the dark.

> NARRATOR
> . . . juvenile haddock between 1985 and 1986, and the numbers of juvenile cod and haddock also dropped dramatically at the same time.

> JOHN CUMMINS (O.S.)
> There never was an investigation into what happened and, and . . .

RESUME — Cummins interview.

> **JOHN CUMMINS**
> . . . what kind of advice the Department gave and, and what influenced the decision-making that went on there. Nothing has ever been . . . you know, it's never been uncovered. It's never been investigated. It's still, you know, a deep, dark secret, what went on and what led to the collapse of the Northern cod in Atlantic Canada.

EXT. ATLANTIC COAST — DAY

A fish farm is located close to shore.

> **NARRATOR**
> The deep, dark secret might not have been a secret at all.

CLOSER — The farm has the distinctive round pens and the black netting covering the pens of an Atlantic Canada fish farm.

Interactions of caligid ectoparasites and juvenile gadids on Georges Bank

John D. Neilson, R. Ian Perry, J. S. Scott, P. Valerio*

Marine Fish Division, Department of Fisheries and Oceans, Biological Station, St. Andrews, New Brunswick E0G 2X0, Canada

ABSTRACT: The role of the ectoparasite *Caligus* sp. (Copepoda: Caligidae) in the northeast Georges Bank cod-haddock ecosystem was examined. Vertical distribution of free-living *Caligus elongatus* adults and host-parasite relationships of juvenile *Caligus* sp. and juvenile gadids are described at 2 locations with contrasting oceanographic properties, one thermally stratified and the other well-mixed. Cod *Gadus morhua* had both greater prevalence and number of *Caligus* sp. ectoparasites than did haddock *Melanogrammus aeglefinus* at both locations. Preferred sites of attachment on the host also differed. While no direct evidence of reduced fish condition as a function of parasite infestation was found, circumstantial evidence is offered in support of the hypothesis that *Caligus* sp. ectoparasitism is a source of mortality for young haddock. Free-living *C. elongatus* were demonstrated to be a significant component of fishes' diet, particularly for cod at the stratified site where zooplankton were less abundant.

INTRODUCTION

The northeast portion of Georges Bank (NW Atlantic) is important for the production of commercially significant gadids, particularly cod *Gadus morhua* and haddock *Melanogrammus aeglefinus*. Surveys of the pelagic stages of O-group gadids conducted by Canadian and United States fisheries agencies have consistently revealed that highest catch rates occur in that vicinity (Cohen et al. 1985, Koeller et al. 1986). In common with other investigations of stocks of gadids world-wide, the Georges Bank studies are often based on the premise that year-class strength is correlated with the extent of natural mortality during the first year of life.

Such work has generally considered predation to be the most important contributor to natural mortality in the first year of life (Sissenwine 1984), and until now parasitism has not received attention. This lack of knowledge is particularly significant given the emerging view that parasitism may have as important a role in the regulation of abundance of natural communities as predation and competition (Dobson & Hudson 1986).

* Present address: Department of Biology, Memorial University of Newfoundland, St. John's, Newfoundland A1B 3X9, Canada

© Inter-Research/Printed in F. R. Germany

In this paper, we examine the role of the ectoparasite *Caligus* sp. (Copepoda: Caligidae) in the northeast Georges Bank cod-haddock ecosystem.

Caligid ectoparasites are known to be damaging to young fish (Rosenthal 1967, Wootten et al. 1982). For example, Kabata (1972) found that the feeding activity of *Caligus clemensi* attached to young pink salmon *Oncorhynchus gorbuscha* resulted in the loss of entire fins. Wootten et al. (1982) revealed that *C. elongatus*, when present on fish in large numbers, can be debilitating even for adult Atlantic salmon *Salmo salar*. Kabata (1974) described the feeding activity of caligid parasites in detail: the fish tissues are scraped off by the strigil (a masticatory apparatus) and the debris is then picked up by the mandibles and conveyed into the buccal cavity. Preliminary histological examination has indicated that the site of attachment and feeding may extend deeper into the fishes' bodies than previously thought (S. MacLean, Oxford Laboratory, National Marine Fisheries Service, Maryland, USA, pers. comm.). As the ectoparasite can be of considerable size (later chalimus stages are about 10 % of the length of the juvenile fish host), indirect effects on the natural mortality of fish might also be expected. Such effects might include increased drag, causing reduced ability to avoid predation, and increased visibility to predators (Holmes & Bethel 1972).

0171-8630/87/0039/0221/$ 03.00

A study led by the Marine Fish Division at the St. Andrews Biological Station discovered unusually high levels of *Caligus elongatus* sea lice. Researchers contemplated whether the samples dominated by sea lice resulted from thermocline conditions in the water column. But this study followed the recent introduction of fish farms to the east coast of Canada and suggests that sea lice populations were expanding exponentially, as they have wherever fish farms have open net pens with thousands of potential hosts in small spaces. (Source: Marine Ecology)

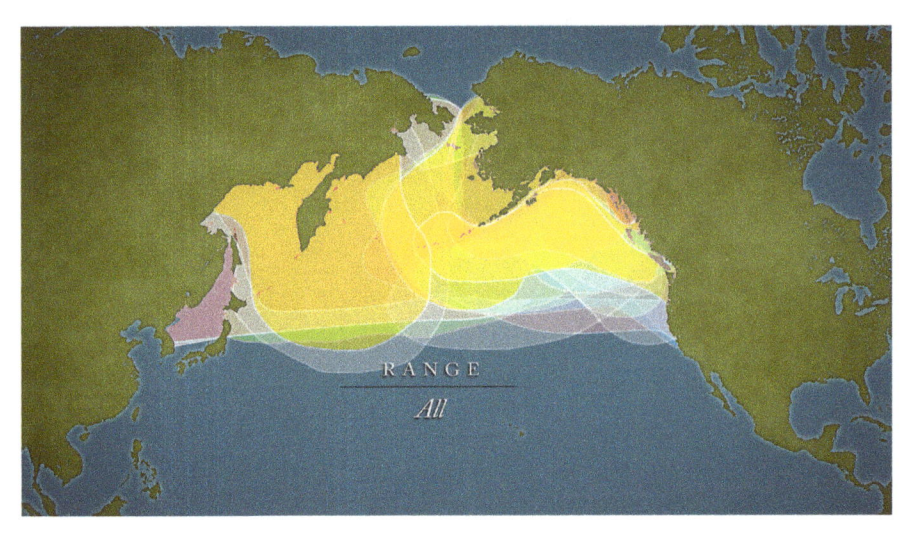

A map of the overlapping ranges of Pacific salmon
species in the North Pacific. (Source: Juggernaut Pictures)

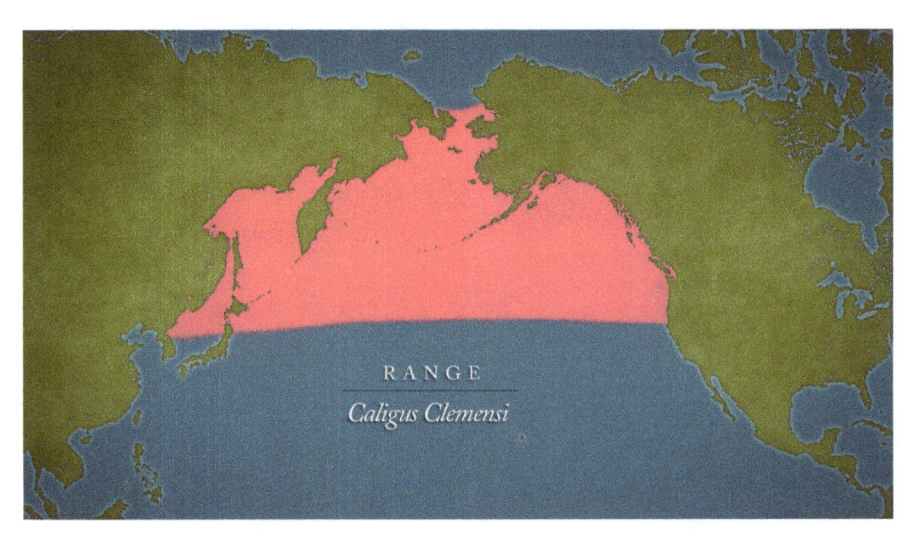

A map of the range of the sea lice *Caligus clemensi* which can parasitize
most species of fish. This means that the parasites can carry and
spread a disease right across the ocean. (Source: Juggernaut Pictures).

Fish farms located on Canada's east coast use a round cage structure and are located in bays and inlets to protect them from rough ocean water and storms. (Photo credit: Anissa Reed)

                    NARRATOR
          It might have been that
          everyone was looking at the
          wrong events.

EXT. ANOTHER EAST COAST FARM — DAY

Seagulls are sitting on the protective net
cover.

                    NARRATOR
          No one suspected that a few
          fish farms could be incubators
          for sea lice and what danger
          that might pose for wild fish
          populations.

INT. HALIFAX HOTEL ROOM — DAY

Alexandra Morton and Inka Milewski are
looking at an Atlantic salmon.

SUPERSCRIPT: Halifax: October, 2012.

                    ANISSA REED (O.S.)
          So, have you heard of problems
          with sea lice this bad?

                    INKA MILEWSKI
          Oh gosh yes. I was, uh . . .
          These, um, this is, this does
          look bad but it's been worse
          actually. Um, there . . . at
                         (MORE)

                    INKA MILEWSKI (CONT'D)
               the height of a sea lice
               epidemic, um, there could be
               much larger open sores on the
               fish. Ah, in fact, sometimes,
               ah, at the worst period they
               uh, they uh, the . . . the
               skull was actually eroded away
               and the face was almost falling
               off.

ANIMATED MAP — The east coast of Canada.

Population status for wild Atlantic salmon
sub-populations is marked as having no
decline, moderate decline and steep decline.

                         NARRATOR
               Wild Atlantic salmon runs
               around the Bay of Fundy are now
               either distressed or extinct
               and cod stocks remain at a
               fraction of what they once
               were.

WILD ATLANTIC FISH GRAPH — One line after
another traces the drop in the populations of
wild fish by species.

                         NARRATOR
               Capelin stocks have virtually
               disappeared, and so have many
               other species.

Inka Milewski removes sea lice from an Atlantic salmon bought at a market. She observed that Atlantic salmon from farms were badly infested with sea lice and reported that many of the farmed fish had half of their faces chewed off by very high sea lice infestations. (Photo credit: Anissa Reed)

EXT. FRASER RIVER — DAY — MORNING

Two First Nations fishermen pull in their net with sockeye.

                    NARRATOR
          On the Pacific coast, evidence
          that sea lice diseases have
          reached far and wide is found
          with a simple abrasion.

OVER THEIR SHOULDERS — They continue pulling in the net.

ON A TOTE — It's full of freshly caught sockeye. The fish on the top have several burn marks on them.

                    NARRATOR
          An abrasion that Fraser River
          First Nations fishermen call
          burn marks.

EXT. HARRISON RIVER — DAY

A sockeye, fresh and dead, has lesions on its side.

                    NARRATOR
          Could these burn marks be the
          same marks described by a DFO
          research scientist in 1981 on
          the east coast?

DOCUMENT COVER — Cone, D.K. "Skin Lesions of Atlantic Salmon." June 1981.

DIAGRAM — The lesions pictured in the appendix of the report.

> NARRATOR
> They were described as mechanical in nature and were found on Atlantic salmon at counting fences near fish farm operations.

A SOCKEYE — It's dead and belly-up in the water. It has a scar on its side.

PHOTO — A dead sockeye with a big scar on its side.

PHOTO — A juvenile sockeye with a louse on it. The scars on its back are from sea lice chewing through the skin to the flesh.

> NARRATOR
> Some of these wounds may be lamprey bites but most lesions and scars are from numerous sea lice . . .

PHOTO — A dead pink salmon spawner, covered in skin lesions.

> NARRATOR
> . . . chewing on the fish throughout its life.

277.

PHOTO — A sockeye covered with multiple fungus patches on its side.

>   NARRATOR
> And even after the fish
> die, fungus grows first where
> the skin is compromised from
> parasite bites creating a
> mottled decomposition process
> on dead and dying salmon.

INT. FULTON RIVER (UNDERWATER) — DAY

Dead sockeye are stacked up like cord wood.

>   NARRATOR
> This is tragic in two ways.
> First of all, many fish that
> have these lesions die before
> spawning. This is called pre-
> spawn mortality and means that
> the disease . . .

EXT. HARRISON RIVER — DAY

Hundreds of sockeye are floating down the Harrison River. They have all died before spawning.

>   NARRATOR
> . . . they were carrying was
> not lethal enough to kill them
> right away, but lethal enough
> to prevent them from completing
> their life cycle.

Skin Lesions of Atlantic Salmon
(Salmo salar) **in Newfoundland Rivers**

D.K. Cone

Research and Resource Services
Department of Fisheries and Oceans
P.O. Box 5667
St. John's, Newfoundland   A1C 5X1

June 1981

Canadian Technical Report of Fisheries and Aquatic Sciences No. 1018

This 1981 study observed skin lesions on 14 wild Atlantic salmon that were described as mechanical in nature. (Source: Government of Canada)

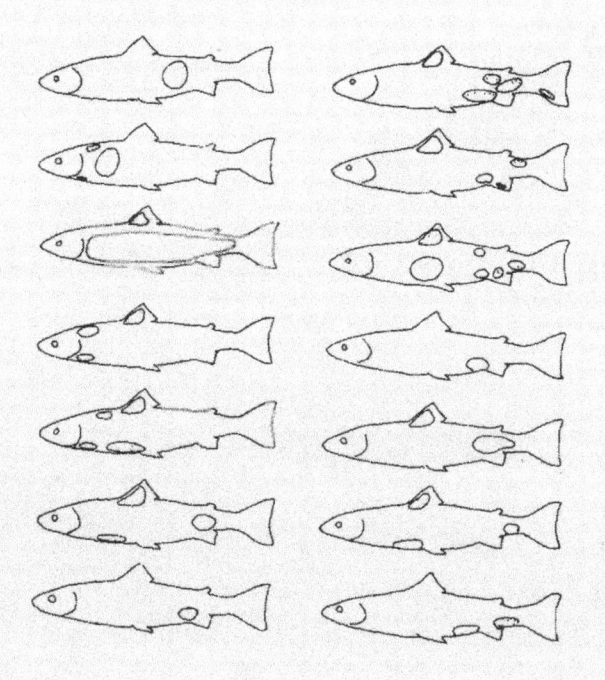

This diagram shows the location of marks on the 14 wild Atlantic salmon observed by J.K. Cone, June 1981. (Source: Government of Canada)

First Nations fishermen noticed what they called burn marks on adult sockeye salmon they caught on the Fraser River in 2011. These burn marks were remarkably similar to marks found on Atlantic salmon on the east coast in 1981. They are caused by large numbers of sea lice chewing on the skin of their host which likely began when the salmon were juveniles. (Photo credit: Scott Renyard)

This is an example of a sea louse chewing on a juvenile salmon. Once the fish matures the marks will look more like lesions. The wounds may not heal because the fish cannot generate new protective scales or mucus, may have been infected with pathogens at that spot, or is not able to prevent other infections from taking hold where the skin is compromised. (Photo credit: Jody Eriksson)

Pacific salmon that have suffered from high sea lice loads during the marine phase of their lives exhibit a mottled fungal infection pattern when they enter fresh water to spawn. Their skin has been compromised and fungus grows in the areas damaged by the sea lice. (Photo credit: Anissa Reed)

EXT. KAKUSHOISH CREEK (2012) — DAY

Thousands of chum salmon are dead.

> NARRATOR
> Second, it means that they are also carrying the disease throughout their range, spreading it to other species that share the marine environment with them.

ANIMATED MAP MONTAGE — The north Pacific fish ranges.

One by one the ranges of the different species are laid over each other.

> NARRATOR
> And because the overlapping ranges of all fish cover the north Pacific, it's also the range for sea lice diseases.

ANIMATED MAP — The north Pacific.

The whole of the north-east Pacific Ocean lights up in red, showing the range of the sea louse *Caligus clemensi*. Then, all the rivers and streams where sea-run fish go extend the impact range of the sea lice. The red colour pulses.

                    NARRATOR
          And it's the same in the north
          Atlantic Ocean.

ANIMATED MAP — The north Atlantic fish
ranges.

Overlapping fish ranges end again with the
overall range for *Caligus elongatus*. And it's
effectively the whole Atlantic Ocean.

                    NARRATOR
          Diseases spread by sea lice
          are likely suppressing fish
          populations right across the
          ocean.

                                        FADE TO:

INT. HATCHERY — DAY

A worker pours eggs into a rearing tank.

                    NARRATOR
          The memos exchanged between
          top scientists and bureaucrats
          in the mid-1980s when Atlantic
          salmon eggs were about to be
          introduced to the Pacific coast
          were fearful of three diseases.

MEMORANDUM — G. Hoskins to M. Comfort RE:
Atlantic salmon eggs from Scotland, December
10, 1985. Page 2. Three diseases are listed
at the bottom of the page.

This map shows that *Caligus elongatus* sea lice have a range that is effectively the range of all Atlantic finfish species combined—that is, the entire north Atlantic Ocean and beyond. (Source: Juggernaut Pictures)

ZOOM IN — Each disease is highlighted.

                    NARRATOR
    Viral hemorrhagic septicemia,
    infectious pancreatic necrosis
    and whirling disease.

EXT. HARRISON RIVER — DAY

At another location downstream, even more sockeye salmon, bloated, are floating next to shore.

ANOTHER SHOT — More dead sockeye.

ON A SINGLE SOCKEYE — Its eye is bulging from its head.

                    NARRATOR
    The clinical symptoms for viral
    hemorrhagic septicemia include
    bulging eyes,

PHOTO — A sockeye juvenile with blood in the eye.

                    NARRATOR
    . . . bleeding in the eyes,
    hemorrhaging in the flesh and
    organs,

PHOTO — A Pacific herring with rose-coloured skin.

                    NARRATOR
            . . . dark skin coloration . . .

PHOTO — A Harrison sockeye with hemorrhaging skin.

                    NARRATOR
            . . . and hemorrhaging in the
            skin.

EXT. RIVER — DAY

A chum salmon is dying in the shallow water near shore. Its skin is yellow and it is obviously suffering from a liver disease.

                    NARRATOR
            But the bad news does not end
            with VHS. Infectious pancreatic
            necrosis has similar symptoms
            but can be easily distinguished
            when fish have . . .

INT. OCEAN — DAY

The sea floor is covered with white squiggly fecal casts over a wide area under a fish farm.

The herring on the left has blood in its eye. This is a typical symptom of a viral hemorrhagic septicemia (VHS) infection. (Photo credit: Anissa Reed)

This sockeye juvenile looks like it is sunburnt, but it is likely suffering from a VHS infection and is bleeding into its tissues. (Photo credit: Jody Eriksson)

NARRATOR
. . . white fecal casts
trailing from their vents. The
sea floor is littered with
white fecal casts under fish
farms and has been seen in wild
fish many times.

TIGHT SHOT — A juvenile fish with a bent spine has whirling disease.

EXT. BEACH — DAY

Mackerel are spinning around and dying.

NARRATOR
Whirling disease, caused by a
pathogenic protozoan, affects
the spine of the fish as it
grows, and it causes fish
to spin in circles, or they
may have black tails. Wild
fish have been seen with this
deformity from the northern
tip of Vancouver Island to
tributaries of the Fraser
River.

FADE TO:

INT. FISH FARM PEN — DAY

Atlantic salmon are packed in a fish farm pen.

Thousands of sockeye salmon died before spawning at Gates Creek, September 9, 2012. This phenomenon is called pre-spawn mortality and is caused by viral infections that become lethal when salmon are more stressed during their migration into fresh water. (Photo credit: Anissa Reed)

Researcher Jody Eriksson displays chum salmon that died before spawning on the Cowichan River. This photograph was taken on November 13, 2012. Many runs on the BC coast involving all Pacific salmon species have experienced large die-offs since the introduction of open net pen fish farms. (Photo credit: Farlyn Campbell)

NARRATOR
The three diseases that everyone feared are now in Pacific coastal waters. Scientists are now saying that these diseases were detected in the 1980s, just a few years after Atlantic salmon were introduced to BC fish farms and about a decade after they were introduced into Puget Sound.

EXT. HARRISON RIVER — MORNING

A silver, but rotting, sockeye salmon is now floating down the river.

NARRATOR
And the most unfortunate part of all is that these diseases are not specific to salmon but can infect any species of fish. And according to the latest studies . . .

INT. HARRISON RIVER — LATER

A silver coho is dead and lying on the bottom.

NARRATOR
. . . viral hemorrhagic septicemia is a highly adaptable virus that infects many fish species.

EXT. HARRISON RIVER — DAY

Several silver sockeye lie dead on a beach.

> NARRATOR
> Its worldwide spread should be considered lethal to all fish.

EXT. ROCKY SHORE — DAY

Two Atlantic salmon lie on the rocks, bleeding from their fins.

> NARRATOR
> And it's now known that Atlantic farmed salmon act as a reservoir and can shed enough virus . . .

INT. PHOTARIUM — DAY

Three herring, bleeding from the fins, are dying.

> NARRATOR
> . . . to cause die-offs in herring approaching 100 percent.

EXT. HARRISON RIVER — DAY

Zipping along in a boat. A few dead salmon float by in the distance.

BLEEDING FINS — Sockeye salmon, Harrison River, September, 2013.

This photo was a key turning point in the making of this film. The bleeding fins changed director Scott Renyard's film from one about the impact of fish farms on wild salmon to one about the impact of fish farms on all wild fish. (Photo credit: Alexandra Morton)

BLEEDING FINS — Rockfish, Sonora Island, July 2013.

BLEEDING FINS — Whitefish, Bulkley River, September 2013.

PRE-SPAWN MORTALITY — Sockeye salmon, Gates Creek, September 2012.

INFECTED LIVER — Sockeye salmon, Adam's River, October 2013.

ABSCESS, YELLOW FLESH, GREEN LIVER — Sockeye salmon, Nass River, September 2013.

DARK SKIN COLORATION — White sturgeon, Fraser River, October 1993.

DARK SKIN COLORATION AND BLEEDING FINS — Sockeye smolt, Blenkinsop Bay, June 2013.

MASS PRE-SPAWN MORTALITY — Pinkut Creek, September 2013.

EXT. HARRISON RIVER — DAY

François, a local fisherman, is standing on the shoreline.

> FRANÇOIS PERREAULT
> Years ago, when I first started coming here which is about three to four years, I used to call this place paradise. Now
> (MORE)

>   FRANÇOIS PERREAULT (CONT'D)
>   it's starting to look more like
>   hell.

HEMORRHAGING SKIN, INFECTED LIVER — Sockeye, Harrison River, September 2011.

PRE-SPAWN MORTALITY — Chum salmon, Cowichan River, November 2013.

PRE-SPAWN MORTALITY — Sockeye salmon, Stellako River, September 2013.

EYE PARASITE — Pacific flounder, Broughton Archipelago, March 2008.

HEMORRHAGING SKIN — White sturgeon, Fraser River, August 1996.

DISTENDED BELLY AND HEMORRHAGING TISSUES — Chum salmon, Kakushoish Creek, September 2012.

FURUNCULOSIS — Pink salmon, Kakushoish Creek, September 2012.

JAUNDICE — Chum salmon, Quarton Creek, September 2012.

BLOOD IN EYE — Pacific herring, Okisollo Channel, June 2012.

BLEEDING FINS — Pacific herring, Malcolm Island, June 2011.

DARK SKIN COLORATION — Pacific herring, Blenkinsop Bay, June 2013.

HEMORRHAGING SKIN, BLEEDING MOUTH AND FINS — Pacific herring, Sointula, August 2013.

EXT. SOINTULA WHARF — DAY

Alexandra Morton holds up a photarium with sick and bleeding herring.

> ALEXANDRA MORTON (V.O.)
> These are juvenile herring caught in the harbour of the fishing village Sointula, off the north coast of Vancouver Island. These fish are very sick. They're bleeding through their scales, they're bleeding in their faces, they're bleeding through their fins. They're dying. There's thousands of them like this. What is this? Is this ok with British Columbia?

FADE TO:

INT. HARRISON RIVER — DAY

A small pea mouth fish is dead on the bottom.

EXT. HARRISON RIVER — DAY

The river is flowing out of Harrison Lake and carrying with it hundreds of silver, dead sockeye.

INT. FULTON RIVER — DAY

Dead sockeye, rotting at the bottom. Hundreds, probably thousands.

                    NARRATOR
          The dramatic drop in fish
          biomass off the coasts of
          Canada, whether they were
          consumed as fish farm food
          or died from diseases, also
          has . . .

INT. ANOTHER STREAM — DAY

Very few salmon. Maybe just one or two.

                    NARRATOR
          . . . serious consequences for
          the environment as a whole.

INT. TRISHA ATWOOD'S LABORATORY — DAY

Establish. Interview. Dr. Trisha Atwood was a PhD candidate at the University of British Columbia when she discovered that carbon dioxide emissions increase dramatically from aquatic ecosystems when predators are removed.

This map highlights where surface layers of the Salish Sea have high concentrations of carbon dioxide. (Source: Juggernaut Pictures)

This world map illustrates the higher levels of carbon dioxide being emitted in the Northern Hemisphere. The generally accepted assumption has been that the concentration of human fossil-fuel-burning activities in the north is the main cause of these higher levels of greenhouse gas emissions. But the tremendous loss of wild fish from disease and the resulting disruption of the marine biological pump is also a major contributor to carbon dioxide emissions. In fact, this may be the single most important cause of global warming since the late 1970s. (Source: Juggernaut Pictures)

TITLE: Dr. Trisha Atwood, Food Web Ecologist, the University of British Columbia.

> TRISHA ATWOOD
> If every organism in an ecosystem contributes in some way to that CO_2 dynamic, it means that large population changes,

INT. OCEAN — DAY

Large numbers of zooplankton and euphasids are hovering.

> TRISHA ATWOOD (O.S.)
> . . . in consumers or primary producers, can really change the amount of respiration or the amount of photosynthesis that is going on within that ecosystem.

INT. OCEAN — DAY

The screen is mostly blank, but green with microscopic phytoplankton.

> TRISHA ATWOOD (O.S.)
> We also know from ecological studies that, um, that things such as trophic cascades which is an ecological phenomenon . . .

INT. WEAVER CREEK — DAY

Chum fry try to attack a leech.

>   TRISHA ATWOOD (O.S.)
>   . . . where the removal or the
>   addition of a predator can
>   actually create these striking
>   changes in the populations of
>   lower trophic levels.

INT. OCEAN — DAY

Herring are milling around feeding.

>   TRISHA ATWOOD (O.S.)
>   And so it was shown that when
>   predators are removed from
>   aquatic ecosystems, this can
>   actually create dramatic
>   changes in the amount of CO_2
>   that is emitted or sequestered
>   from an ecosystem.

EXT. SALISH SEA — DAY

The greenhouse fog over the ocean blocks our view of Vancouver Island.

>   PIETER TANS (O.S.)
>   The atmosphere is always, you
>   could say, beholden or slave to
>   what ocean carbon is doing.

INT. PIETER TANS'S LAB — DAY

Establish. Interview.

TITLE: Dr. Pieter Tans, Senior Scientist, NOAA Earth System Research Laboratory.

> PIETER TANS
> There's 40 times more carbon in the oceans than there is in the atmosphere. So on a very long time scale the oceans rule. They determine how much will be in the atmosphere.

EXT. PACIFIC COAST (AERIAL) — DAY

The Pacific coast stretches off far into the distance.

> PIETER TANS (O.S.)
> But the atmosphere doesn't feel the entire oceans. It feel, feels only the surface of the oceans and that's where biology comes in.

INT. OCEAN — DAY

Many small fish are schooling around under a structure.

                    TRISHA ATWOOD (O.S.)
               When a fish eats a plant, a
               part of that plant is actually
               accrued into the biomass of the
               fish and other parts of it is
               respired out as CO_2.

INT. OCEAN - DAY

A spectacular view of the sea floor with a
wide array of algae and plants.

                    TRISHA ATWOOD (O.S.)
               And so that's how carbon can
               be fixed within an ecosystem.
               So it goes into either biomass
               of animals or it can go into
               biomass of plants.

EXT. BURRARD INLET - DAY

Looking across the surface, then slowly
sinking below the surface.

                    PIETER TANS (O.S.)
               With ocean organisms, they,
               they take carbon out of surface
               waters and photosynthesize them
               into organic material, lowering
               CO_2 in surface waters.

DIAGRAM - Shows the biological pump in cross-
section. Part one is where the arrows are
pointing into the ocean.

                    TRISHA ATWOOD
          If the CO_2 in the water is
          much lower than that of the
          atmosphere, then it tries
          to take up CO_2. So it starts
          becoming a CO_2 sink.

DIAGRAM — Now CO_2 is being released into the
atmosphere.

                    TRISHA ATWOOD (O.S.)
          And so, if there's too much
          CO_2 in the water, if it's much
          greater than the amount of
          CO_2 that's in the overlying
          atmosphere then it tries to
          emit CO_2.

INT. OCEAN — DAY

The surface waters distort the sky.

                    NARRATOR
          Carbon dioxide is normally pH-
          neutral until it mixes with
          water and forms carbonic acid.
          So acidic ocean water is an
          indication that it contains
          high levels of carbon dioxide.

EXT. SHORELINE — DAY

Mussels cling to the rocks.

NARRATOR
Recent reports have found that shellfish farms located in the Salish Sea have been experiencing die-offs due to unusually acidic ocean conditions.

ANIMATED MAP — BC's south coast.

A fog appears over the Salish Sea.

NARRATOR
This means that the surface waters of the Salish Sea contain high levels of CO_2 and are now pumping greenhouse gases into the atmosphere off the BC coast.

RETURN TO THE COVER — "A Brief on Mariculture," 1972. A page turn takes us to page 2.

NARRATOR
The ultimate irony is the first report on aquaculture in 1972 identified . . .

THE WORDS — "The coastal waters of B.C. are among the most productive (high carbon-fixation rate) of the North Pacific" are highlighted.

                    NARRATOR
          . . . "The coastal waters of BC
          are among the most productive
          of the North Pacific."

PANNING ACROSS — a large fish farm with
seagulls hovering over the nets.

                    NARRATOR
          But by trying to use that
          productivity for growing fish
          in pens,

ON THE WATER — Sunlight reflects off the
ocean surface.

                    NARRATOR
          . . . this same ecosystem is no
          longer fixing carbon like it
          used to and is . . .

EXT. SALISH SEA — DAY

A bright sun glares into the camera.

                    NARRATOR
          . . . instead emitting CO_2 into
          the atmosphere. And this is
          probably happening wherever
          there are . . .

EXT. PACIFIC COAST FISH FARM — DAY

A ten-pen fish farm in the middle of a bay.

                    NARRATOR
          . . . open net pen fish
          farms affecting wild fish
          populations.

                    PIETER TANS (O.S.)
          We have been measuring carbon
          dioxide and . . .

RESUME — Tans interview.

                    PIETER TANS
          . . . other greenhouse gases
          for about four decades, all
          over the world. And, ah, from
          that data, we can see how CO_2
          and these other greenhouse
          gases are rising in the world
          everywhere. And we also see
          spatial patterns. We see, for
          example, that most of these
          greenhouse gases are higher in
          the Northern Hemisphere than in
          the Southern Hemisphere.

ANIMATED MAP — of the world.

The carbon dioxide fog spreads over the top
half of the world.

                    NARRATOR
          In fact, the Northern
          Hemisphere emits eight times
          more CO_2 than the Southern
                         (MORE)

                    NARRATOR (CONT'D)
          Hemisphere. Much of that
          difference can be explained
          by the fact that more fossil-
          fuel-burning humans live in
          the north. But how much of that
          difference is due to impaired
          biological pumps in the north
          Pacific and Atlantic oceans?

RESUME — Tans interview.

                    PIETER TANS
          When modern carbon dioxide
          measurements started in the
          late 1950s,

NOAA'S GRAPH — It shows the trend line of CO_2 over time in ppm.

                    PIETER TANS (O.S.)
          . . . the average annual rate
          of increase was 0.7 parts per
          million per year, and now it's
          more than two parts per million
          per year,

RESUME — Tans interview.

                    PIETER TANS (O.S.)
          . . . so it's a threefold
          increase in the rate at
          which we're pumping these
          things into the atmosphere.

EXT. PACIFIC COAST — DAY

A foggy channel. The sunshine is trying to break through.

					NARRATOR
			It might just be an interesting
			coincidence that atmospheric
			CO_2 has escalated since fish
			farms were placed in the oceans
			in the early 1970s. But maybe
			it's not a coincidence at all.

EXT. PACIFIC COAST (AERIAL) — DAY

Back to the two fish farms from the opening.

					NARRATOR
			So if we ponder that earlier
			question, how many fish farms
			would it take to be a problem?
			One farm.

INT. PACIFIC BIOLOGICAL STATION NET PEN — DAY

Chum salmon are milling around. And we can see sea lice on the back of one of them.

					NARRATOR
			And in that farm you would only
			need one fish, either wild
			or farmed, to have a virulent
			pathogen. And by providing
					(MORE)

                    NARRATOR (CONT'D)
          abundant hosts, it can rapidly
          reproduce and spread quickly
          into the environment.

EXT. ADAM'S RIVER — DAY

The red back of a healthy sockeye salmon sticks out of the shallow water.

                    NARRATOR
          And within a very short time,
          it becomes a very large problem.

EXT. SALISH SEA — DAY

A windy Salish Sea with clouds on the horizon.

                    NARRATOR
          There have been three
          significant events that have
          led to a much warmer planet.

EXT. FARM — DAY

Establish.

                    NARRATOR
          The first was the clearing of
          land for settlement and food
          production.

EXT. PORT MANN BRIDGE — DAY

The highway is full of cars.

                    NARRATOR
          The second was the burning of
          fossil fuels.

INT. OCEAN — DAY

A shot of consumers.

EXT. RIVER — DAY

Sockeye salmon are trying to spawn.

                    NARRATOR
          And now the destruction of the
          biological pumps in the ocean.

INT. OCEAN — DAY

There are no plants. No life at the bottom.

                    NARRATOR
          Will this last event be the one
          to end mankind's run on this
          planet?

                                        FADE OUT:

FADE IN:

EXT. VICTORIA STREET — DAY

We pick up the protest. The protesters are now walking through the streets of Victoria.

TIGHTER — The crowd streams by, waving signs.

SUPERSCRIPT: Get Out Migration — Protest Day 16.

SONG: UP YOUR WATERSHED

>   KEVIN WRIGHT (V.O.)
>   Red backs cruisin' up the
>   Fraser River.

A HORSE BUGGY — goes by as part of the protest.

>   KEVIN WRIGHT (V.O.)
>   And the pinks are doin' it too!

THE CROWD — is getting larger.

>   HOLLY ARNTZEN AND KEVIN WRIGHT (V.O.)
>   Find their way up river
>   valleys, shooting through those
>   grizzly alleys. Salmon runnin',
>   I hope they're comin' soon.

ALEXANDRA — is waving a flag in the direction of the crowd.

    HOLLY ARNTZEN AND KEVIN WRIGHT (V.O.)
      Up your watershed!

EXT. ISLAND HIGHWAY — DAY

The protest makes its way down the hill.

    HOLLY ARNTZEN AND KEVIN WRIGHT (V.O.)
      What's been happenin' in
      the ocean. There musta been
      some big commotion. Are they
      fashionably late or comin' at
      all? Up your watershed. Up your
      watershed!

        BARTLETT NAYLOR (O.S.)
      Even as an outsider I can see
      and I've begun to appreciate
      how the first maxim of
      ecology . . .

EXT. STREET — DAY

The Elder who appeared at the beginning of the film passes through the frame.

        BARTLETT NAYLOR (O.S.)
      . . . that everything is
      connected is no truer than it
      is for salmon.

Protesters walk down Highway 19 on the outskirts of Victoria, BC, May 8, 2010. (Photo credit: John Preston)

The Get Out Migration protest grows in size as it enters the city of Victoria, BC. (Photo credit: John Preston)

People stream onto the lawn of the BC legislature on May 8, 2010. (Photo credit: Jack Cooley)

On May 8, 2010, a crowd filled the lawn of the legislature and pressed toward the steps to hear several speakers, including Joan Phillip, Grand Chief Stewart Phillip, Rafe Mair, Vicky Husband, Brian Gunn, Fin Donnelly, Thea Block, Chief Bob Chamberlin, Billy Proctor and of course, Alexandra Morton. (Photo credit: John Preston)

RESUME — Naylor interview.

>   BARTLETT NAYLOR
> I don't know what British
> Columbia would be like without
> salmon. It just wouldn't be
> British Columbia.

EXT. BRITISH COLUMBIA LEGISLATURE — DAY

Off the building, we find the protest.

>   HOLLY ARNTZEN AND KEVIN WRIGHT (V.O.)
> Up your watershed! Up your
> watershed! Up your watershed!
> Woo!

The crowd appears to be bopping to the beat.

>   HOLLY ARNTZEN (V.O.)
> Up your watershed! Woo, woo,
> woo!

EXT. CROWD — DAY

Lots of protesters milling around.

BACK TO THE MICROPHONE — Chief Darren Blaney is speaking.

>   DARREN BLANEY
> Long time ago when humans first
> came to this earth, they were
> so pitiful. And all the animals
>   (MORE)

DARREN BLANEY (CONT'D)
and all the creatures, the plants, they got together and started to see how can we help the pitiful humans? And they asked, "Who's going to help the humans?" The first one to stand up was the salmon.

THE CROWD — cheers.

BACK TO BLANEY

DARREN BLANEY
And as the salmon stood up, all the other creatures of the ocean also stood up to help the humans. People, in return, were supposed to look after all of the things of this earth. And I think we've lost our way. Just as our governments have lost their way.

EXT. LAWN OF THE LEGISLATURE — DAY

We find Alexandra in the crowd. She looks exhausted, but smiles. The people of BC have come out to fight for wild salmon. Will it be enough?

FADE TO:

SUPERSCRIPT: Just a few months after the Get Out Migration protest, the Canadian government struck a "Commission of Inquiry" into the Decline of the Fraser River Sockeye.

SUPERSCRIPT: As of 2014, fish farms continue to operate off the coasts of Canada. The Canadian Government plans to expand the industry.

EXT. PRISTINE BC COAST — EVENING

Pretty sparkling water.

EXT. PRISTINE COAST (AERIAL) — DAY

Beautiful coastal landscape.

EXT. ROCKY SHORELINE — DAY

Waves crash spectacularly against the rocks.

INT. OCEAN — DAY

A magnificent school of herring.

EXT. STREAM — DAY

A Kermode bear fishes for pink salmon in a stunningly pretty stream.

EXT. INCH CREEK — DAY

A shot of the resident sturgeon. Many fish swim around it.

Alexandra Morton speaks at the final rally in Victoria. She is happy that so many supported the protest and spoke up for BC's wild salmon. (Photo credit: John Preston)

INT. OCEAN — DAY

A sea lion leaves a trail of bubbles behind it.

INT. ADAMS RIVER — DAY

An impressive school of bright red sockeye.

INT. OCEAN — DAY

Sea anemones, bright with colour, are hanging from rocks.

EXT. STREAM — DAY

Pink salmon fins stick out of the water.

EXT. GREAT BEAR RAINFOREST — DAY

Two bear cubs are on a log. One white and one black.

INT. STREAM — DAY

Four newly hatched alevins gulp their first breaths.

INT. STREAM — DAY

Juvenile coho swim in crystal clear water.

INT. OCEAN — DAY

Herring swim in unison.

INT. OCEAN — DAY

A school of cod.

INT. ADAM'S RIVER — DAY

Sockeye salmon, bright red, but a normal red, are vying for space on the spawning beds.

EXT. OCEAN — DAY

Immature herring swim by in the shallows.

EXT. CHANNEL — DAY

A pristine channel. No farms.

                                    FADE OUT:

ROLL TAIL CREDITS

SONG: GET OUT MIGRATION.

                    THE END

Acknowledgements

The *Pristine Coast* was a taxing and challenging project that would not have been possible without the support and expertise of many people. I was suffering from the ravages of Lyme disease throughout the project, and it's highly likely this film would never have seen the light of day had I been working alone.

I am eternally grateful to all of the interview subjects—Geoff Meggs, John Allen Fraser, John Cummins, Otto Langer, Stephanie Elderton, Jody Eriksson, Megan Adams, Darren Blaney, Bob Chamberlin, Alejandro Buschmann Rubio, Jack Cooley, Ernie Crey, Robert Fraumani, Thor Froslev, Greg McDade, Alexandra Morton, Bartlett Naylor, Craig Orr, Daniel Pauly, William Proctor, June Quipp, Krista Robertson, Don Staniford, Rashid Sumaila, John Volpe, Peter H. Pearse, Brad Hope and June Hope, and Rafe Mair—who gave generously of their time. As with all of my films, not all the interviews made the final cut, but the knowledge I gained from all the interviews played a key role in my understanding of the subject and made the film so much better.

As documentarians know, our films are told with a very small and dedicated crew. This film was no different, and I would like to sincerely thank my DP, Mark Noda, who managed tight windows for interviews and other time crunches that came at us daily. Filming the Get Out Migration involved many ferry and road trips to catch up with the protest as it moved down Vancouver Island. Once we were at the various locations, we often had to sprint down a street or road to catch the protest as it was unfolding. I open the film with one of those sprints when we accidentally left the camera on. It was a lucky mistake that gave us a wonderful opening moment.

My editor, Maja Zdanowski, was amazing at extracting the emotional core from the footage. Her skill at editing is highlighted in particular by the disease montage sequence, which will bring tears to your eyes.

The film is also blessed with the musical talents of Richard Hagensen, Joanne Banks, Holly Arntzen, and Kevin Wright, who shared their terrific songs. They brought the energy of the film to a world-class level, and their songs revealed the deep emotional connection British Columbians have to wild salmon.

I would also like to thank the many suppliers who provided images and footage for the film: Twyla Roscovich, Damien Gillis, Dick Harvey, Kathleen Harvey, Helen Slinger, Jeff Turner, Jody Eriksson, Marvin Rosenau, Tavish Campbell, Alexandra Morton, Anissa Reed, Brad Hope, June Hope, Zachery Pinsent, Folk Ryden, Richard Davies, Trevor Amos, Tim Foulkes, Robert Michelson, Raymond Plourde, Douglas Hay, The Fisherman Publishing Society, The Conservation Fund, Freshwater Institute, and the Vancouver Aquarium.

I would like to make a special mention of two BC environmental film makers who are no longer with us: Dick Harvey and Twyla Roscovich. Dick worked for the Department of Fisheries and Oceans, and made early environmental films like *Living River* and *Indian Food Fishing on the Fraser River*. His dedication to Pacific fish and wildlife was captured over several decades, and I have drawn on his library of material several times. He is sorely missed. The lovely and talented Twyla made many short films and was incredibly dedicated to her craft and the BC environment. She passed away after the making of this film. BC has lost a major force in the world of film and a wonderful person. I miss her and her storytelling immensely.

A huge thank-you goes to Brad and June Hope, whose candid interviews about the early fish farm days in BC were essential to the making of this film. There is no substitute for first-hand accounts when it comes to real-world events. And having a better understanding of the evolution of the aquaculture industry is critical if we are to address its impacts on wild fish populations.

Special thanks to The Green Channel Inc. (thegreenchannel.tv) for all the work you do to promote the protection of wild fish populations.

A huge thanks to Jan Westendorp for her amazing skill and guidance in managing the design and publishing of this book. I could not have done it without her. And to Lesley Cameron, whose terrific and prompt story editing kept the book project moving forward and a dream project to work on.

Special thanks go to my family for putting up with my crazy film-making schedules and the many days I was off filming somewhere. Not to mention the days I spend staring blankly into space while I try to find a word or better way to tell a story.

Lastly, I would like to dedicate this book to my father, Thomas Basil George Renyard, who passed away on November 1, 2021. His constant encouragement in my early years pushed me to find out more about the plants and animals around me by going to university in the big city of Vancouver. My background in science, and eventually in film, gave me the tools to investigate this story in much greater depth, and I was able to draw conclusions that broke new ground in evaluating the impacts of aquaculture in the Northern Hemisphere. My findings are thanks to you, Dad.

Bibliography

ARTICLES: JOURNALS

Beamish, R.J., Neville, C.M., Sweeting, R.M., Jones, S.R.M., Ambers, N., Gordon, E.K., Hunter, K.L., and McDonald, T.E. (2006, December). A proposed life history strategy for the salmon louse, *Lepeophtheirus salmonis* in the subarctic Pacific. *Aquaculture, 264,* pp. 428–440.

Caine, G. (1988, May). Measures to ensure safe fish farming in British Columbia. *Aquaculture Information Bulletin,* 25.

Ford, J.S., and Meyers, R.A. (2008). A global assessment of salmon aquaculture impacts on wild salmonids. *PLOS Biology, 6*(2), e33.

Gottesfeld, A.S., Proctor, B., Rolston, L.D., and Carr-Harris, C. (2009). Sea lice, *Lepeophtheirus salmonis*, transfer between wild sympatric adult and juvenile salmon on the north coast of British Columbia, Canada. *Journal of Fish Diseases, 32,* pp. 45–57.

Hicks, Brad. (2012, January/February). Anatomy of deception. *Aquaculture North America,* p. 5.

Hutchings, J. A., Walters, C., and Haedrich, R. L. (1997). Is scientific inquiry incompatible with government information control? *Canadian Journal of Fisheries and Aquatic Sciences, 54,* pp. 1198–1210.

Krkosek, M., Ford, J.S., Morton, A., Lele, S., Myers, R.A., and Lewis, M.A. (2007, December). Declining wild salmon populations in relation to parasites from farm salmon. *Science, 18,* pp. 1772–1775.

Krkosek, M., Lewis, M.A., Morton, A., Frazer, L.N., and Volpe, J.P. (2006, October). Epizootics of wild fish induced by farm fish. *PNAS, 103*, pp. 15506–15510.

Krkosek, M., Lewis, M.A., Volpe, J.P., and Morton, A. (2006). Fish farms and sea lice infestations of wild juvenile salmon in the Broughton Archipelago—A rebuttal to Brooks (2005). *Reviews in Fisheries Science, 14*, pp. 1–11.

Morton, A. (2005, December). Mortality rates for juvenile pink *Oncorhynchus gorbuscha* and chum *O. keta* salmon infested with sea lice *Lepeophtheirus salmonis* in the Broughton Archipelago. *Alaska Fishery Research Bulletin, 11*(2), pp. 146–152.

Morton, A., et al. (2004). Sea lice (*Lepeophtheirus salmonis*) infection rates on juvenile pink (*Oncorhychus gorbuscha*) and chum (*Oncorhynchus keta*) salmon in the nearshore marine environment of British Columbia, Canada. *Canadian Journal of Fisheries and Aquatic Science, 61*, pp. 147–157.

Morton, A., Routledge, R., and Krkosek, M. (2008). Sea louse infestation in wild juvenile salmon and Pacific herring associated with fish farms off the east-central coast of Vancouver Island, British Columbia. *North American Journal of Fisheries Management, 28*, pp. 523–532.

Morton, A., and Symonds, H.K. (2002). Displacement of *Orcinus orca* (L.) by high amplitude sound in British Columbia, Canada. International Council for the Exploration of the Sea. *Journal of Marine Science, 59*, pp. 71–80.

Morton, A., and Volpe, J. (2002). A description of escaped farmed Atlantic salmon *salmo salar* captures and their characteristics in one Pacific Salmon Fishery Area in British Columbia, Canada, in 2000. *Alaska Fishery Research Bulletin, 9*(2), pp. 102–110.

Morton, A., Routledge, R.D., and Williams, R. (2005). Temporal patterns of sea louse infestation on wild Pacific salmon in relation to the fallowing of Atlantic salmon farms. *North American Journal of Fisheries Management, 25*, pp. 811–821.

Nyland, A., Hovland, T., Hodneland, K., Nilsen, F., and Lovik, P. (1994). Mechanisms for transmission of infectious salmon anaemia (ISA). *Diseases of Aquatic Organisms, 19*, pp. 95–100.

Roth, M. (1993, May). Tools of the trade. *Northern Aquaculture.*

Roth, M. (1993, May–June). The future of sea lice control in Canada: To treat or not to treat? *Northern Aquaculture.*

Todd, C.D., Walker, A.M., Ritchie, M.G., Graves, J.A., and Walker, A.F. (2004, September). Population genetic differentiation of sea lice (*Lepeophtheirus salmonis*) parasitic on Atlantic and Pacific salmonoids: Analyses of microsatellite DNA variation among wild and farmed hosts. *Canadian Journal of Fisheries and Aquatic Sciences, 61*(7), pp. 1176–1190.

ARTICLES: The *Fisherman* newspaper (published by The Fisherman Publishing Society), with no named author

$500,000 study focuses on impacts of salmon farming. (August 28, 1989).

1.6 million pink catch forecast. (June 21, 1976).

10 million sockeye. (December 11, 1981).

14 million sockeye headed for Fraser. (December 12, 1985).

1984 salmon regulations: DFO tightens economic screws. (April 19, 1984).

1985 Fraser forecasts: 16m pinks, 9m sockeye. (December 12, 1984).

1993 net fisheries forecasts of Pacific salmon catch. (May 24, 1993).

1998 escapement highest level ever for Fraser run. (September 28, 1998).

1999 tidal water fishing zones north coast. (June 28, 1999).

20 million pink run forecast for the Fraser. (December 10, 1982).

350,000 farm salmon escape as winter storm smashes pens. (February 17, 1989).

93 Sites in Works—Gold rush mentality hits fish farms. (July 19, 1985).

Abundance of Pacific herring stocks Strait of Georgia. (February 24, 1997).
Aiding salmon spawners: Improvements proceed. (August 26, 1977).
Alaska bill would ban salmon farms. (March 25, 1988).
Alaska fishermen mobilize to prohibit salmon farming. (December 12, 1985).
Alaska ready to remove farm ban. (March 25, 1987).
Alaska runs falling short. (August 15, 1986).
Alaska sockeye down 15% as Bristol Bay winds up. (August 22, 1994).
Alaska troll sector fights farmed salmon impact. (August 15, 1986).
Alcan exempted from review. (October 22, 1990).
Alcan seeks Kemano approval. (January 20, 1984).
Algae hits salmon farms. (July 18, 1986).
Algae, disease, low prices hammer B.C. salmon farmers. (September 25, 1989).
All's not well down on the farm. (August 16, 1985).
America's shame at Bristol Bay. (April 23, 1965).
Antibiotic use out of control, scientist warns fish farmers. (September 16, 1988).
Applauds survival coalition: Fraser stands by election pledges. (January 18, 1985).
Aquaculturalist supports Pearse's approach. (January 14, 1983).
Aquaculture. (June 23, 1972).
Aquaculture: 2. (May 18, 1984).
Aquaculture industry seeks full development subsidies. (November 16, 1984).
Aquaculture: Lummi Indians rear coho, trout. (June 23, 1972).
Aquaculture smolts prefer to stay home. (December 10, 1982).
Aquaculture the boom out of control. (June 23, 1986).
Are herring also doomed? (April 23, 1965).
Atlantic salmon site on Kennedy Lake opposed by UFAWU and sports fishers. (April 20, 1992).

B.C. fishermen see destruction of wild stocks in Norway. (April 22, 1988).
Barkley fishery is so poor fishermen pull nets early. (June 10, 1983).
Better management needed for Central Coast. (January 21, 1991).
Big business—getting bigger. (May 14, 1965).
Big catches made by herring fleet. (March 10, 1972).
Big companies rule aquaculture in B.C. (September 21, 1992).
Big sockeye landings in Johnstone Strait. (August 23, 1993).
Bristol Bay catches cut in half as run collapses. (July 21, 1997).
Bristol Bay run worst since 1897. (September 1, 1972).
Bristol Bay runs weak, prices strong. (July 18, 1986).
Bristol Bay sockeye catch down by 20%. (July 25, 1994).
Bristol Bay sockeye catch second biggest on record. (July 20, 1990).
Bristol Bay's sockeye below 1997 "disaster." (July 20, 1998).
Bristol run fails. (July 21, 1997).
Call for compensation. (August 23, 1999).
Cameron, J. "There's tremendous potential fish production out of these streams." (December 20, 1993).
Campbell River guides take aim at Johnstone strait chum fishery. (March 21, 1994).
Canfisco plans seen as key to industry future. (March 15, 1999).
Cannery row. (December 17, 1990).
Catch limit retained for Atlantic herring. (March 16, 1973).
Catches jump for second opening on Fraser River. (August 19, 1996).
Change in SEP. (September 21, 1984).
Chile's farm salmon thrive on low wages. (December 14, 1992).
Chlorine killed Capilano salmon. (November 10, 1975).
Chums come on strong. (November 23, 1998).
Claim of $220 million fish aid misleads public. (April 27, 1998).
Coast communities demand farm freeze. (December 12, 1986).
Coho crisis excerpts from Copes report. (May 25, 1998).
Coho escapement trends 1953–1993. (March 23, 1998).

Coho made a comeback on Skeena, stats show. (April 15, 1999).
Columbia has heavy spring salmon run. (November 30, 1973).
Comeback on the cod. (October 21, 1996).
Commercial Fraser sockeye catch set at half 1991 level in poor cycle year. (February 24, 1992).
Common Property: Private ownership leads to extinction of stocks. (July 20, 1990).
Consumer group urges farm salmon controls. (May 23, 1989).
Co-operation among all users vital to save salmon resources. (November 23, 1992).
Crosbie says no to jobs. (November 18, 1991).
Current policies to import Atlantic salmon eggs amount to an act of "biological insanity." (April 18, 1986).
Danger at the farm. (April 18, 1986).
Deadly virus hits salmon on fish farm. (November 23, 1992).
DFO acted without regard to the public interest, judge rules. (December 16, 1996).
DFO admits goof on Goldstream chums. (December 12, 1985).
DFO aquaculture subsidy hits $3 million in '85. (July 18, 1986).
DFO baffled by Barkley sockeye, chinook returns. (August 21, 1987).
DFO calls for ban on Skeena gillnets to help steelhead. (November 18, 1991).
DFO changed stand on fish farm pact. (September 18, 1987).
DFO gives the green light to more foreign egg imports. (November 18, 1985).
DFO isolates new exotic disease in farm Atlantics. (December 13, 1989).
DFO memo slams farm chaos. (December 12, 1986).
DFO releases farm chinook. (June 23, 1989).
DFO renewing implementation of Pearse Report. (January 19, 1990).
DFO salmon expectations for 1989. (February 17, 1989).
DFO says no restrictions on dumping disease fish. (January 22, 1988).
DFO sees west coast closure, more areas in '85 roe fishery. (April 19, 1984).

Disease up at Scottish fish farms. (May 14, 1990).

Diseases threatens Norway's farms. (September 18, 1987).

Disputes over 1994 repayments, travel stamp add to UI problems. (October 23, 1995).

Drastic Bristol Bay closure confirmed for 1959 season. (May 8, 1959).

Duane Holberg. (March 18, 1991).

Early Stuart sockeye down from forecast. (July 26, 1995).

Early Stuart sockeye escapement. (August 24, 1992).

EEC herring sales could jump. (September 8, 1978).

Eggs from 9 chinook stocks taken for farmers' gene pool. (November 20, 1987).

Egmont farmers set up plant. (June 23, 1986).

EI eligibility for fishermen will be based on income. (July 22, 1996).

End to fish farm freeze sparks new protest, anger. (March 25, 1987).

Escape of Atlantics the 'largest ever' in B.C. fish farms. (August 23, 1993).

Escapements hit record as fishermen by-passed. Vol. 63 No. 9 (September 28, 1998).

Escapements okay despite PSC miscount. (January 24, 1996).

Escapements show coho rebounding. (December 18, 1998).

Farm fish escape near Kyuquot. (January 21, 1991).

Farm fish to hit wild salmon market. (June 23, 1986).

Farm salmon glut set to depress markets. (January 25, 1989).

Farmed, Russian salmon changing market outlook. (June 30, 1997).

Farmed salmon fails when HOFA's counted. (May 22, 1987).

Farmers hit with fines for illegal mort dumps. (May 23, 1989).

Farmers using SEP facility. (March 17, 1989).

Farmers, fishermen united in opposition to free trade. (January 22, 1988).

Farms hit more stormy weather. (August 23, 1993).

Farms offer 'surplus' chinook. (May 23, 1989).

Farms threaten key gillnet fishery. (June 21, 1993).

First caught since 1833: Lone salmon in Thames. (December 18, 1974).

Fish farm agreement nearing completion. (July 17, 1987).
Fish farm diseases killing wild stocks in Norway. (January 22, 1988).
Fish farm drug hits wild stock. (January 25, 1989).
Fish farm genetic program robs common property chinook stocks. (November 20, 1987).
Fish farm moratorium sought in Pender. (July 18, 1986).
Fish farm policy sparks concern for wild stocks. (October 25, 1999).
Fish farm problems mount in both Norway and B.C. (September 24, 1990).
Fish farmers avoid issue on vhs find. (April 21, 1989).
Fish farmers hit regulations to protect wild salmon stocks. (April 19, 1985).
Fish farmers try sockeye despite disease threat. (March 16, 1990).
Fish farmers versus fishermen. (April 16, 1987).
Fish farms: A bad idea getting worse. (August 16, 1985).
Fish farms breed problems for communities, wild stock. (June 21, 1985).
Fish farms continue to face opposition and salmon kills. (September 21, 1992).
Fish farms create new problems. (August 16, 1985).
Fish farms eyed by Nova Scotia. (April 17, 1973).
Fish farms in ussr. (September 17, 1965).
Fish farms use of antibiotics poses threat. (December 13, 1988).
Fish industry ills threaten sep gains. (August 17, 1979).
Fish versus aluminum. (March 16, 1984).
Fisheries delays new regulations. (April 12, 1979).
Fishermen demand herring changes for 1986 season. (December 12, 1985).
Fishing closures, big run buffers put record numbers on grounds. (December 16, 1996).
Fishing industry major force in B.C. economy, study finds. (October 24, 1988).
Fleet praised for coho conservation. (November 23, 1998).
Fleet takes first cuts. (July 22, 1996).

Food fishery is still an issue on Fraser. (October 25, 1999).

Food multinationals are taking over B.C. salmon farms. (March 16, 1990).

Forecast shows two strong years. (April 21, 1989).

Fraser board dam sites. (March 20, 1959).

Fraser crisis real, says Pearse probe. (December 14, 1992).

Fraser pledges co-operation in aquaculture development. (March 15, 1985).

Fraser River salmon management plan. (July 22, 1991).

Fraser River sockeye will be reasonably good in 1990. (May 14, 1990).

Fraser run at 4.2 million. (August 19, 1996).

Fraser SEP is "impossible" under Tory budget plans. (March 15, 1985).

Fraser sockeye catch now set at 10 million. (May 26, 1997).

Fraser water quality good... But toxic pollution increasing. (January 16, 1976).

Fraser's demise reveals real Tory fisheries policy. (October 18, 1985).

Free trade: Tory strategy threatens sovereignty, jobs and fishing industry. (March 21, 1986).

Full public review needed of Alcan's Kemano II project. (December 17, 1990).

Gillepsie lifts farm freeze but major problems unresolved. (January 16, 1987).

Gillnet roe herring fishery poor in Gulf: seines take West Coast quota in two days. (March 16, 1990).

Gillnetter Terry Lubzinski with Atlantic caught in straights fishery in 1996 [photo caption]. Fish and Ships. (August 18, 1987).

Gillnetters shocked at haul of farm fish. (October 24, 1988).

Good year led by strong Fraser River stocks. (April 24, 1990).

Governments, farm industry undertake first checks on quality, antibiotic use. (February 17, 1989).

Green light given for offshore drilling. (May 21, 1986).

GVRD sewage case adjourned. (September 25, 1995).

'Halt aquaculture leases' —union. (February 21, 1986).

Herring fishery uproar. (November 18, 1985).
Herring landing largest since sixties. (January 28, 1977).
Herring on high: Gulf Stocks at high level belying consultants' claim. (February 24, 1997).
Herring spawn 388 miles in 1974. (January 24, 1975).
Herring spawn exceeded 1973 record last spring. (February 13, 1976).
Herring stocks at '64 level on entire south, central coast. (October 18, 1985).
Herring: The story of a squandered resource. (December 19, 1975).
Hidden agenda: DFO wants boat quotas and licence limitation. (October 22, 1990).
Higher Skeena run forecast: Sockeye run set at 2.6 million. (January 25, 1974).
Horsefly run: Once famous sockeye run makes comeback. (September 9, 1977).
Horsefly sockeye run a "miracle." (August 16, 1985).
Howe Sound moratorium freezes salmon farmers. (July 18, 1986).
Huge Skeena run spawns uproar. (October 18, 1985).
Illegal dumping of farmed fish is widespread. (April 21, 1989).
Indians slam farm boom. (October 24, 1986).
Ireland freezes foreign fish farms. (July 18, 1986).
Irish fishermen manage herring. (November 17, 1978).
Islands demand fish farm freeze. (October 10, 1986).
Japanese tagged sockeye found in Johnston strait. (January 10, 1975).
Just what was Bastien saying on aquaculture? (December 10, 1999).
Kemano 2 plans will decimate Fraser sockeye. (December 14, 1979).
Low returns create poor season in 1992. (August 24, 1992).
Maritime Weston affiliate begins salmon farming. (May 21, 1986).
Meggs steps down as editor after 12 years of service. (April 24, 1990).
Memos bare farm's secrecy request. (January 24, 1996).
Mixed signals for '85: Processors reap sockeye bonanza. (May 17, 1985).

Moratorium threatened: Green light given for offshore drilling. (May 21, 1986).
More protection for herring sought by Pender Harbour. (February 13, 1959).
Most fish farm production controlled by foreign firms. (April 16, 1987).
Nanaimo salmon tank farm would be illegal in Norway. (May 22, 1987).
N.B. salmon men battle Davis. (March 16, 1973).
NDP affirms stand on fish farm freeze. (May 23, 1989).
NDP urges salmon farm licensing freeze. (June 17, 1988).
NDP vows farm freeze, coastal plan. (June 17, 1988).
New algae bloom hits Gulf salmon farmers. (May 22, 1987).
New salmon farming leases would harm Sabine fishery. (October 24, 1986).
New vaccine for salmon. (February 24, 1975).
No fisheries on Fraser sockeye, industry urges. (March 18, 1996).
North coast salmon canneries: A pictorial history of the early salmon canning industry. (December 17, 1990).
Northern pink return a mystery. (July 29, 1988).
Norway fears farm salmon releases. (July 17, 1987).
Norway moves on wild stock as BC leases explode. (September 19, 1986).
Norway sees mounting disease, pollution threat from fish farms. (December 13, 1988).
Norway slaps ban on herring fishing. (January 10, 1975).
Norway tour finds wild stock disaster. (April 22, 1988).
Norway's salmon farms face tight regulation. (June 23, 1986).
Offshore drilling inquiry 'shapes up as a farce.' (August 17, 1984).
Offshore drilling opponents hammer Chevron on spills. (October 18, 1985).
Oil drilling, supertankers threaten fishery resource. (January 30, 1981).
Ombudsman report urges new Aquaculture law, coast zones. (January 25, 1989).

Optimism wanes as fish runs falter. (July 19, 1985).
Ottawa-Victoria memo seen as 'a step forward.' (July 22, 1996).
Owikeno returns at record levels. (November 17, 1986).
Owikeno stream survey finds disappointing sockeye returns. (November 17, 1989).
Partial ban ordered on pit-lamping herring. (August 6, 1965).
Pattison offering $13m for Canfisco. (January 20, 1984).
Pearse opposition (continued from page 1). (December 10, 1982).
Pender Harbour fishers don't want licensing changes that will hurt coastal towns. (May 21, 1991).
Photo of accelerated coho at Pacific Biological Station rearing pens. (October 29, 1976).
Pickets target farm fish. (August 19, 1988).
Pink forecast 3.3 million: Skeena sockeye prediction 1.3 million. (January 12, 1977).
Plants gear for work as first openings set. (June 21, 1993).
Pollution causing decline. (September 29, 1972).
Poor season improved with good fall chum catches in Johnstone Strait. (October 19, 1992).
Premier halts fish farms. (November 17, 1986).
Private hatcheries 'on agenda.' (March 25, 1987).
Private, for profit: Ocean ranching—Oregon fishermen say it is wiping them out. British Columbia is next on the private ocean ranchers' agenda. (March 15, 1985).
Privatization, Pearse still on DFO agenda. (June 21, 1985).
Processors may yield UI in tariff fight. (March 21, 1986).
Proposed Fraser River training walls project. (October 4, 1977).
Protest against aquaculture forces Victoria policy review. (June 23, 1986).
Province to demand TBT tests on salmon. (May 22, 1987).
Province, DFO to probe TBT contamination from fish farms. (December 12, 1986).

Puget seiners land Atlantics. (October 24, 1988).
'Rebuild Fraser salmon runs.' (February 23, 1988).
Report claims Statscan plays down job crisis. (September 8, 1978).
'Restore freeze on farms'—UFAWU. (January 16, 1987).
Rivers, Smith inlets targeted for fish farms. (July 19, 1993).
Royal Pacific salmon farm empire in receivership. (November 17, 1989).
Salmon area filled (continued from page 1). (April 8, 1977).
Salmon commission sets forecast for Fraser runs. (April 21, 1989).
Salmon enhancement: Public hearings prove fishermen were right. (April 8, 1977).
Salmon farm escapes demand DFO action. (February 17, 1989).
Salmon farmers denounce 'fabrications' on disease. (July 15, 1988).
Salmon farming now faces full review. (April 24, 1995).
Salmon farms pollute Fundy. (November 16, 1988).
Salmon return in record numbers, prices hit bottom. (September 27, 1993).
Salmon stocks could double. (December 13, 1988).
Salmon study probes human health links. (September 24, 1990).
Salmon treaty. (July 28, 1999).
Salmon welfare fund sorely needs injection. (June 15, 1962).
Savage permits 200 new farms. (January 16, 1987).
Scots seek N.S. herring. (November 17, 1978).
Scottish fish farms organize. (April 24, 1990).
Scottish fishermen claim oil drilling destroys fishing. (September 20, 1985).
Scottish herring closure legacy of overfishing. (September 8, 1978).
Scottish salmon face farm salmon threat. (July 25, 1989).
Scraping bottom: Herring stocks are in trouble but DFO can't say why or when recovery might begin. (November 18, 1985).
SEP funds needed for Fraser runs. (September 20, 1985).
Short roe herring season close to quota. (March 25, 1977).
Siddon to give Province control over aquaculture. (April 18, 1986).

Siddon urges Pearse consultative plan. (September 19, 1986).
Signals mixed from herring fishery. (March 23, 1998).
Skeena coho 'at risk' without treaty fishing agreement with Alaska. (July 20, 1998).
Skeena fisheries again corked by harvest plan. (July 22, 1996).
Skeena runs threatened: Disease hits Fulton sockeye fry. (August 17, 1984).
Skeena sockeye run 2.4 million. (September 7, 1973).
So what's a sockeye equivalent? (June 24, 1991).
Sockeye catch reaches 4.8 million. (September 20, 1974).
Sockeye catches hold up well in several areas. (July 27, 1973).
Sockeye forecast for Fraser down. (December 18, 1974).
Sockeye plug Adams River. (October 24, 1986).
Sockeye Review Board Report. (March 20, 1995).
Solteau only fisherman on aquaculture council. (July 17, 1987).
Some new, some old regulations set to apply for EI fishing claims. (October 21, 1996).
Some salmon farms denied in Smith and Rivers Inlets. (May 23, 1994).
South coast forecast for 1991–1994. (June 24, 1991).
South coast net forecast for 1991–1994. (February 15, 1991).
Spill of farmed Atlantics highlights control issue. (December 13, 1994).
Stand up for Canada, UFAWU tells Ottawa. (March 18, 1996).
Steelhead in crisis: who is to blame? (April 22, 1991).
Stop whale export until family taken. (July 16, 1965).
Sunshine Coast puts brakes on aquaculture development. (January 17, 1986).
Sunshine Coast retains salmon farm freeze. (January 16, 1987).
Support grows for aquaculture freeze. (December 12, 1986).
Surplus farm smolts dumped in Jervis. (July 25, 1989).
Surprise pink return in north baffles DFO. (August 10, 1983).
T Buck Suzuki Environmental Foundation. (January 22, 1988).
TBT banned for fish farms after B.C. supply seized. (March 25, 1987).

The complete text of the Canada-B.C. fisheries agreement. (April 28, 1997).

The foreshore rush. (June 23, 1986).

The government record: When the fishing industry has asked for assistance, the Social Credit cabinet always has refused. (October 10, 1986).

The politics of numbers. (November 23, 1994).

The Salmonid Enhancement Program: It produces fish, creates jobs and pays economic dividends—but can it survive? (September 21, 1984).

Three million pink return to Johnstone Strait seen. (July 29, 1977).

Tories hire 700 to hit UI users. (December 12, 1984).

Total Fraser runs of eight million sockeye forecast. (December 17, 1973).

Tough new UI provision won't kick in until 1997. (June 24, 1996).

Tragedy on the Vedder. (March 6, 1964).

Treaty 'failure': Anderson unable to get U.S. to curb coho catch. (June 29, 1998).

Trend is to aquaculture. (December 18, 1974).

UFAWU assails herring fishery irregularities. (January 28, 1977).

UFAWU calls for assistance as Fraser sockeye run falls. (August 21, 1995).

UFAWU certified at AgriMarine. (March 18, 1996).

UFAWU claims victory over UI changes. (May 22, 1987).

UFAWU demands better herring management. (December 17, 1990).

UFAWU outlines policy in brief to review. (March 24, 1997).

UFAWU pressing for changes to Skeena sockeye fishing plan (continued from page 1). (July 22, 1996).

UFAWU, salmon farmers debate license freeze. (December 12, 1986).

UFAWU sets 'balanced policy' on fish farming. (March 23, 1998).

UFAWU voices fishermen's concern: Herring fishery mismanaged. (April 21, 1978).

UI targeting fishing industry workers. (January 19, 1994).

Union demands aquaculture pact meeting. (May 22, 1987).

Union seeks halt to fish farm law. (March 25, 1988).

Union to continue efforts for joint action by trollers. (April 28, 1972).

Unsolved riddles haunt salmon plans. (July 18, 1986).

Upper Adams sockeye return called 'miracle': Good ocean survival, reduced fishing credited with big jump in escapement. (September 23, 1996).

US group wants Fraser fish. (April 22, 1991).

US sockeye half million ahead. (July 27, 1973).

U.S. trade action hits B.C. fishery. (July 18, 1986).

Use UI aid to re-build. (September 25, 1995).

Vedder River project ruins chum spawn. (March 6, 1964).

Vigil on the Adams: Legendary sockeye run faces a renewed threat warns the Adams River Society. (June 24, 1996).

Virus threatens wild salmon, steelhead. (March 17, 1989).

'Wage controls must go', labor tells Trudeau gov't. (August 26, 1977).

Waiting for answers: Trio's hunger strike prompts meetings with federal official. (October 25, 1999).

'War on fishermen': Total closure on Fraser. (January 30, 1981).

Warning of stock decline, DFO slashes '85 roe quotas. (October 19, 1984).

Water reform nets record Calif. run. (November 20, 1995).

Weaver Creek sockeye likely around 200,000. (October 19, 1998).

West coast herring "in trouble." (April 22, 1982).

What's happening to the coho? (March 23, 1998).

Whose debt? (September 25, 1995).

Wild salmon conference looks to future. (October 24, 1988).

Young salmon, herring are farm food. (April 21, 1989).

ARTICLES: The *Fisherman* newspaper (published by The Fisherman Publishing Society), with named author

Bartosh, G. (1996, July 22). Fraser coho target of new study.

Brodowich, B. (1999, October 25). Sockeye closure creating hardship in communities.

Brown, D. (1994, November 23). Can Tobin talk of conservation while decimating DFO's ranks?

Brown, D. (1994, December 13). Fishing industry at the crossroad—A brief presented to the Fraser River Sockeye Review Board by the UFAWU.

Cameron, J. (1994, November 23). Pared-down Owikeno spawning survey points to new uncertainty in the future.

Drouin, M. (1993, December 20). Jervis Inlet: Story and photos.

Griffin, S. (1994, January 19). RCMP probe of Kemano cites Ottawa interference.

Griffin, S. (1994, July 25). Alexandra Morton: A watchful eye on salmon farming.

Griffin, S. (1995, September 25). Keeping a fishery.

Griffin, S. (1995, December 18). Changing weather: Shift in patterns having a dramatic effect on salmon, groundfish stocks.

Griffin, S. (1998, December 18). Voices of hope from Echo Bay.

Griffin, S. (1999, July 28). Weather changes could benefit marine survival.

Lane, D. (1995, February 20). Is there a future for our fisheries?

Meggs, G. (1984, April 19). Aquaculture: Fledgling industry offers peril and promise.

Morton, A. (1994, July 25). 'We need answers, not stock phrases.'

Morton, A. and B Proctor. (1998, December 18). Excerpts from *Heart of the Raincoast: A Life Story*, in the *Fisherman*.

Ransome, T. (1994, September 27). Halt to fish farms urged.

Articles: Miscellaneous newspapers

Douthwaite, R. (July 7, 1990). Anglers fear fish-farms in West are killing wild salmon. *Sunday Tribune*.

Grigg, R. (April 16, 2010). Alexandra's great "Get Out Migration." *Campbell River Courier*.

Judson, N. (January 9, 2009). Streamkeeper warns of salmon fish farm, salmon advocate educates residents at BAG as part of Eagle Festival. *The Chief.*

Judson, N. (January 23, 2009). The Salmon Whisperer: Leading expert Alexandra Morton urges action against fish farms. *The Chief.*

Pynn, L. (February 5, 2009). First Nations use of salmon. *Vancouver Sun.*

The Georgia Straight. (December 3–10). Mystery on the river.

Books

Fort, C., Daniel, K., and Thompson, M. (January 2009). *Herring spawning survey manual.* Fisheries and Oceans Canada Science.

Groot, C., and Margolis, L. (1991). *Pacific salmon life histories.* UBC Press.

Reports

Amos, K.H., and Appleby, A. (1999, September). *Atlantic salmon in Washington State: A fish management perspective.* Washington Department of Fish and Wildlife.

Anderson, A.D. (1983, September). *The migration and exploitation of chum salmon stocks of the Johnstone Strait—Fraser River Study Area, 1962–1970.* Canadian Technical Report of Fisheries and Aquatic Sciences No. 1166.

Anderson, B., Castledine, A., Kieser, D., Ludwig, B., Reid, G., and Hohnsbehn, C. (1993, July 16). *Meeting of the Transplant Committee.* Department of Fisheries and Oceans, Government of Canada.

Aro, K.V. (1979, March). *Transfers of eggs and young of Pacific salmon within British Columbia.* Fisheries and Marine Service Technical Report No. 861. Government of Canada.

Austin, R. (2006, February). *Special Committee on Sustainable Aquaculture* (Report of the Proceedings). Hansard. Province of BC.

Ayles, G.B., and Brett, J.R. (1978, January). *Workshop on aquaculture research and development in Canada.* Fisheries and Marine Service Technical Report No. 750. Government of Canada.

Barber, F.G. (1983, May). *Inshore migration of adult Fraser sockeye, a speculation.* Canadian Technical Report of Fisheries and Aquatic Sciences 1162.

BC Government. (1997, April 11). *Policy for the importation of Atlantic/Pacific salmon into British Columbia.* BC Ministry of Environment.

Beacham, T.D. (1982, February). *Some aspects of growth, Canadian exploitation, and stock identification of Atlantic cod* (Gadus morhua) *on the Scotian Shelf and Georges Bank in the Northwest Atlantic Ocean.* Canadian Technical Report of Fisheries and Aquatic Sciences 1069. Government of Canada.

Brett, J.R., Griffioen, W., and Solmie, A. (1978, November). *The 1977 crop of salmon reared on the Pacific Biological Station experimental fish farm.* Fisheries and Marine Service Technical Report 845.

Brett, J.R., Calsprice, J.R., Ghelardi, R.J., Kennedy, W.A., Quayle, D.B., and Shoop, C.T. (1972, February). *A brief on mariculture.* **Fisheries Research Board of Canada, Technical Report No. 301.**

Cone, D.K. (1981, June). *Skin lesions of Atlantic salmon* (Salmo salar) *in Newfoundland rivers.* **Canadian Technical Report of Fisheries and Aquatic Sciences No. 1018.**

Conley, D.C. (1994, February 17–18). Kudoa workshop proceedings. Province of British Columbia, No. 94–01.

Gibson, R.J. (1981, September). *Behavioural interactions between coho salmon* (Oncorhynchus kisutch), *Atlantic salmon* (Salmo salar), *brook trout* (Salvelinus fontinalis), *and steelhead trout* (Salmo gairdneri) *at the juvenile fluviatile stages.* **Canadian Technical Report of Fisheries and Aquatic Sciences No. 1029.**

Gilhousen, P. (1990). *Prespawning mortalities of sockeye salmon in the Fraser River system and possible causal factors.* International Pacific Salmon Commission, Bulletin XXVI.

Government of Canada. (n.d.) DFO *Pacific Region position on the U.S. response to the Atlantic salmon importation protocol.*

Government of Canada. (2011). COSEWIC *assessment and status report on the Atlantic salmon* Salmo salar. COSEWIC Secretariat, Canadian Wildlife Service.

Hourston, A.S. (1978, June). *The decline and recovery of Canada's Pacific herring stocks.* Fisheries and Marine Service. Technical Report 784.

Hourston, A.S., and Outram, D.N. (1972, January). *Millions of eggs and miles of spawn in British Columbia herring spawnings, 1951 to 1970.* Fisheries Research Board of Canada. Technical Report No. 296.

Hyatt, K., Johannes, M.S., and Stockwell, M. (2007). *Ecosystem overview: Pacific North Coast Integrated Management Area (PNCIMA).* Canadian Technical Report of Fisheries and Aquatic Sciences 2667.

Kennedy, W.A. (1974, March). *Sablefish culture—Final report.* Fisheries Research Board of Canada. Technical Report No. 452.

Kennedy, W.A. (1977, June). *The 1976 crop of salmon reared on the Pacific Biological Station experimental fishfarm.* Fisheries and Marine Service Technical Report No. 726.

Kennedy, W.A., and Smith, M.S. (1973, May). *Sablefish culture—progress in 1972.* Fisheries Research Board of Canada. Technical Report No 388.

Kieser, D., Traxler, G.S., and Evelyn T.P.T. (c. 1995). *Isolation of viral haemorrhagic septicemia (vhs) virus from herring* (Clupea harengus pallas) *in British Columbia. Pacific Biological Station.* [Draft technical report].

Kohler, A.C., Faber, D.J., and McFarlane. N.J. (1974). *Eggs, larvae, and juveniles of fishes from plankton collections in the Gulf of St. Lawrence during 1965, 1966 and 1967.* Fisheries Research Board of Canada. Technical Report No. 285.

Krkosek, M., Gottsfeld, A., Proctor, B., Rolston, D., Carr-Harris, C., and Lewis, M.A. (2007). Effects of host migration, diversity and aquaculture on sea lice threats to Pacific salmon populations. Proceedings of the Royal Society, 274, 3141–3149.

Labelle, M. (2009, July). *Status of Pacific salmon resources in Southern British Columbia and the Fraser River Basin*. Pacific Fisheries Resource Conservation Council.

Lightly, D.T., Wood, M.J., and Heizer, S.R. (1985, July). *The status of chum salmon stocks of the west coast of Vancouver Island, 1951–1982, Statistical Areas 22–27*. Canadian Technical Report of Fisheries and Aquatic Sciences No. 1366.

MacCrimmon, H.R., Stewart, J.E., and Brett, J.R. (1974). Aquaculture in Canada: The practice and the promise. ***Bulletin of the Fisheries Research Board of Canada*, Bulletin 188.**

Macdonald, D.S. (1985). *Royal Commission on the Economic Union and Development Prospects for Canada*. Volume one. Government of Canada.

May, A.W., and Lear, W.H. (1971, August). *Digest of Canadian Atlantic salmon catch statistics*. Fisheries Research Board of Canada. Technical Report No. 270. McDade, G.J., Glowacki, L., and Morton, A. (2011, December 29). *Supplementary argument—ISA hearings. Aquaculture Coalition Exhibit for The Uncertain Future of Fraser River Sockeye*. Government of Canada Inquiry.

Narver, D.W. (1972, July). *A survey of possible effects of logging on two eastern Vancouver Island Streams*. Fisheries Research Board of Canada. Technical Report No. 323.

Neilson, J.D., Perry, R.I., Scott, J.S., and Valerio, P. (1987, September 10). ***Interactions of caligid ectoparasites and juvenile gadids on Georges Bank*. Marine Ecology—Progress Series V, 39: 221–232. Government of Canada.**

Noakes, D.J., Beamish, R.J., and Gregory, R. (2002, September). *British Columbia's commercial salmon industry*. North Pacific Anadromous Fish Commission.

Orr, C., Stewart, C., and Price, M. (2009, February). *Environmental groups commend Salmon Forum's acknowledgement of fish farm problems, demand government action on key recommendations*. Coastal Alliance for Aquaculture Reform.

Pearse, P.H. (1982, September). *Turning the tide: A new policy for Canada's Pacific fisheries.* The Commission on Pacific Fisheries Policy. Final Report.

Price, I., and Nickum, J.G. (1991, July 17). *Summary report.* FTA Technical Working Group on Fish and Fishery Products Inspection.

Proctor, B. (1996). *Decline of chum salmon to Shoal Harbour Creek graph, Scot Cove Hatchery report.* [Unpublished report].

Province of British Columbia. (2011, July). *Policy and practice report: Aquaculture regulation in British Columbia.*

Ricker, W.E., Bilton, H.T., and Aro, K.V. (1978, September). *Causes of the decrease in size of pink salmon* (Onchorhychus gorbuscha). Fisheries and Marine Service Technical Report 820.

Saxby, D.J., Filion, L., L'Aventure, J., Mackean, R., Post, G., and Robbins, N. (1984, August). *Aquaculture: A development plan for Canada.* Final report of the Industry Task Force on Aquaculture for The Science Council of Canada.

Sibert, J., and Parker, R.R. (1972, June). *A model of juvenile pink salmon growth in the estuary.* Fisheries Research Board of Canada. Technical Report No. 321.

Stroomer, C., and Wilson, M. (2013, January). *British Columbia's fisheries and aquaculture sector.* 2013 Edition for Department of Fisheries and Oceans. BC Stats.

Sutterlin, A.M., Harmon, P., and Barchard, H. (1976). *The culture of brook trout in salt water,* Fisheries and Marine Service. Technical Report No. 636.

Taylor, F.H. (1973, November). *Herring tagging experiments, 1957–67.* Fisheries Research Board of Canada Technical Report No. 422.

Verhoeven, L.A., and Davidoff, E.B. (1962). *Marine tagging of Fraser River sockeye salmon.* International Pacific Salmon Fisheries Commission, Bulletin XIII: 1–132.

Volpe, J. (2001, October). *Super un-natural—Atlantic salmon in BC waters.* Report for The David Suzuki Foundation.

Wahle, R.J., and Pearson, R.E. (1987, September). *A listing of Pacific Coast spawning streams and hatcheries producing chinook and coho salmon with estimates on numbers of spawners and data on hatchery releases.* NOAA Technical Memorandum NMFS/NWC-122.

Wildsmith, B.H. (1984, April). *Federal aquaculture regulation.* Canadian Technical Report of Fisheries and Aquatic Sciences No. 1252.

Withler, F.C. (1982, April). *Transplanting Pacific salmon.* Canadian Technical Report of Fisheries and Aquatic Sciences No. 1079.

Letters

Thank you to Alexandra Morton for providing me with copies of all the letters listed below. They comprise copies of the letters she wrote, copies of responses to her letters, and letters she acquired via a Freedom of Information request.

Anderson, D. Letter from D. Anderson to B. Proctor. August 7, 1998.

Anderson, D. Letter from D. Anderson to A. Morton, Raincoast Research at Simoon Sound RE: No evidence. February 23, 1998.

Anderson, D. Letter from D. Anderson to A. Morton, Raincoast Research at Simoom Sound. RE: Doing all we can. February 9, 1999.

Anderson, D. Letter from D. Anderson to A. Morton, Raincoast Research at Simoom Sound. RE: Impacts. July 5, 1999.

Anderson, D. Letter from D. Anderson to A. Morton, Raincoast Research at Simoom Sound. RE: Accept that fact. May 10, 1999.

Anderson, D. Letter from D. Anderson to A. Morton, Raincoast Research at Simoom Sound. September 4, 1998.

Andrusak, H. Letter to A. Morton, Raincoast Research. RE: Response to March 29, 1993, letter. April 16, 1993.

Andrusak, H. Letter from H. Andrusak, Manager Fish Culture Section to A. Morton. July 7, 1993.

Andrusak, H. Letter from H. Andrusak, Manager Fish Culture Section to A. Morton. RE: More research. June 15, 1993.

Andrusak, H. Letter from H. Andrusak, Manager Fish Culture Section to A. Morton, Raincoast Research at Simoom Sound. RE: Escape sitings. June 30, 1993.

Andrusak, H. Letter from H. Andrusak, Manager Fish Culture Section to A. Morton. RE: Impacts. March 16, 1993.

BC Environment Minister. Letter from BC Environment Minister to A. Morton. RE: Placement of tenures. June 10, 1993.

BC Ministry of Environment. Letter from Ministry of Environment to A. Morton, Raincoast Research at Simoom Sound. January 19, 1996.

Bell Irving, R. Letter to F.D. Austin Pelton, Minister of the Environment from the Steelhead Society of British Columbia. RE: Importation of Atlantic salmon eggs. December 28, 1985.

Berry, A. Letter from A. Berry to A. Morton, Raincoast Research at Simoom Sound. February 23, 1995.

Berry, A. Letter from A. Berry to A. Morton, Raincoast Research at Simoom Sound. February 28, 1997.

Berry, D. Letter from D. Berry, Land Officer to A. Morton, Raincoast Research at Simoom Sound. March 15, 1993.

Berry, D. Letter from D.W. Berry, Land Officer to A. Morton, Raincoast Research at Simoom Sound. RE: Grandfathered tenure. May 10, 1994.

Berry, D. Letter from D.W. Berry, Land Officer to A. Morton, Raincoast Research at Simoom Sound. RE: Opposing fish farm sites. May 27, 1993.

Berry, D. Letter from D.W. Berry, Land Office to A. Morton, Raincoast Research at Simoom Sound. August 10, 1994.

Berry, D.W. Letter to A. Morton. RE: Response to July 19, 1993. August 11, 1993.

Berry, D.W. Letter from D.W. Berry, Land Officer to A. Morton, Raincoast Research at Simoom Sound. RE: Opposing fish farm sites. June 4, 1993.

Blackburn, D. Letter to D. Narver, Director, Fisheries Branch BC Ministry of Environment and Parks. RE: Restrictions placed on Atlantic salmon egg importation. October 23, 1986.

Bouchard, J. Letter from J. Bouchard to A. Morton, Raincoast Research Society. RE: Sea lice issue not accepted completely. December 5, 2003.

Boutillier, J. Letter to A. Morton, Director, Raincoast Research. RE: Response to February 1, 1993, letter. February 9, 1993.

Caine, G. Letter to A. Morton, Raincoast Research from BC Ministry of Agriculture, Fisheries and Food. RE: Fish farm placement in the Broughton Archepelago. July 5, 1993.

Carey, T.G. Letter to G.I. Pritchard. RE: Importation of cultured Atlantic salmon from Norway, Aquaculture and Resource Development Branch, Department of Fisheries and Ocean, Government of Canada. December 10, 1982.

Cashore, J. Letter to Honourable Bill Barlee, Minister of Agriculture, Fisheries and Food. RE: Disease outbreak of IHN (Infectious Hematopoetic Necrosis). October 28, 1992.

Chamut, P. Letter to Dave Narver. RE: Consolidating federal/provincial policy on the importation of Pacific salmon into British Columbia to include trout. June 10, 1988.

Chamut, P. Letter to S. Culbertson, Assistant Deputy Minister, Fisheries and Food Division, Ministry of Agriculture, Fisheries and Food. RE: New shellfish aquaculture policy. August 12, 1993.

Chamut, P. Letter to Washington Fish Growers Association. RE: Response to letter of July 13, 1992. August 18, 1992.

Chamut, P. Letter to B.C. Salmon Farmers Association. RE: Quarantine and disease policies. August 21, 1991.

Chamut, P. Letter to A. Morton, Raincoast Research. RE: several letters about salmon farming in the Broughton Archipelago. August 25, 1993.

Chamut, P. Letter to A. Morton, Raincoast Research. RE: Antibiotic resistant *Aeromonas salmonicida* in Broughton archipelago fish farms. December 9, 1993.
Chamut, P. Letter to Alexandra Morton. December 9, 1993.
Chamut, P. Letter from P. Chamut, Director General of Pacific Region to A. Morton, Raincoast Research at Simoom Sound. RE: Outbreaks of furunculosis in salmon cultured in the Broughton Archipelago. February 14, 1994.
Chamut, P. Letter to B.A. Hackett, Assistant Deputy Minister, Ministry of Agriculture and Fisheries. RE: Federal—Provincial policy on the importation of live salmonids into British Columbia. February 3, 1987.
Chamut, P. Letter to B.J. Emberley, Director General, Inspection Services Directorate. RE: Atlantic salmon importation policy. February 6, 1992.
Chamut, P. Letter to G. Barrows, Free Trade Co-ordination Division, External Affairs and International Trade. RE: Restrictions limiting the importation of Atlantic salmon into British Columbia. January 23, 1991.
Chamut, P. Letter to J. Lawrie. RE: Importation of Atlantic salmon eggs. July, 1990.
Chamut, P. Letter from P. Chamut, Director General Pacific Region Fisheries & Oceans to A. Morton, Director Raincoast Research at Simoom Sound. RE: Presentation. May 7, 1993.
Chamut, P. Letter from Pat Chamut, Director General Fisheries Pacific to [redacted fish farmer]. November 14, 1985.
Chamut, P. Letter to [redacted company]. RE: Importation of Atlantic salmon eggs from New Brunswick. November 25, 1991.
Chamut, P. Letter to Sea Farm Canada Inc. RE: Importation of Atlantic salmon eggs from New Brunswick. November 25, 1991.
Chamut, P. Letter to R. Schmitten, Regional Director, U.S. Department of Commerce, National Marine Fisheries Service. RE: US importation of Atlantic salmon eggs meeting. November 29, 1991.

Chamut, P. Letter to J. Mckay, United Hatcheries Ltd. RE: Importation of Atlantic salmon eggs. October 2, 1992.

Chamut, P. Letter from P. Chamut, Director General of Pacific Region to J. McKay, United Hatcheries Ltd. RE: Importation of Atlantic salmon eggs. [undated].

Cull, E. Letter from E. Cull, Minister to A. Morton, Raincoast Research at Simoom Sound. RE: Broughton a study case. August 15, 1995.

Cull, E. Letter from E. Cull, Minister to A. Morton, Raincoast Research at Simoom Sound. RE: Disease transfer. July 21, 1995.

Cull, E. Letter from E. Cull, Minister to A. Morton, Raincoast Research at Simoom Sound. June 2, 1995.

Cull, E. Letter from E. Cull, Minister to A. Morton, Raincoast Research at Simoom Sound. June 28, 1995.

Cummins, J. Letter from J. Cummins to Senator F. Murowski, United States Senate. March 7, 2002.

Davies, J. Letter from J. Davies, Professor Emeritus, Department of Microbiology and Immunology at University of B.C. to A. Morton. March 25, 1999.

Davis, J. Letter from J.C. Davis, Regional Director-Science of Pacific Region to A. Morton, Raincoast Research at Simoom Sound. RE: Scott Cove. January 22, 1998.

Davis, J. Letter from J.C. Davis, Regional Director-Science to A. Morton, Raincoast Research at Simoom Sound. RE: Seal deterrent. May 8, 1996.

Deans, D.L. Letter from D.L. Deans, Director Habitat Management at Pacific Region, DFO to A. Morton at Rainbow Research. RE: Habitat in the Broughton archipelago. July 6, 1993.

Dhaliwal, H. Letter from H. Dhaliwal to A. Morton. RE: Marine mammals. August 16, 2000.

Dhaliwal, H. Letter from H. Dhaliwal to A. Morton, Raincoast Research at Simoom Sound. RE: Several issues. November 24, 2000.

Dicks, N. Letter to T. J. Billy. RE: Lifting of Canadian ban on the importation of US produced salmonid seed stocks. June 11, 1991.

Douglas Jr., J.E. Letter to J. Emberley, Inspection, Regulations, and Enforcement. January 03, 1992.

Drinkwater, K. Letter from K. Drinkwater to A. Morton, Raincoast Research at Simoom Sound. RE: Algal blooms. December 19, 1994.

Emberley, B.J. Letter to James Douglas, Director, Office of Trade and Industry, NOAA, US Department of Commerce. February 14, 1992.

Emberley, B.J., and T.J. Billy. Letter to J. Nickum and I. Price, Co-chairpersons of the Sub-working Group for Aquaculture and Fish Diseases. June 19, 1991.

Fleming, I. Letter from I. Fleming, Norwegian Institute for Nature Research to A. Morton, Raincoast Research at Simoom Sound. [undated].

Forester, J. Letter to Thomas Billy, Director Trade and Industry Services Division NOAA from Washington Fish Growers Association. RE: Canada's policy on egg imports from the US. May 2, 1990.

Gibson, J. Letter from J. Gibson to A. Morton. November 6, 1993.

Ginetz, R. Letter to British Columbia Trout and Char Producers. RE: Trout and char importation policy. January 16, 1990.

Ginetz, R. Letter to D. Narver, Director Fisheries Branch, Ministry of Environment and Parks. RE: 20,000 eggs/licence/year import restriction. April 12, 1988.

Ginetz, R. Letter from Ron Ginetz to P. Chamut, Director General of Pacific Region, Department of Fisheries and Oceans. RE: Policy of importation of salmonids. April 21, 1987.

Ginetz, R. Letter from R. Ginetz, Regional Aquaculture Coordinator of Pacific Region to A. Morton, Raincoast Research at Simoom Sound. RE: Acoustic deterrent device usage for aquaculture predation control. August 2, 2000.

Ginetz, R. Letter to D. Narver. RE: Importation of eyed Atlantic ova into BC for broodstock development purposes. August 31, 1990.

Ginetz, R. Letter from R.M.J. Ginetz to Dr. F. Forgeron, Philips Arm Sea Farms Ltd. March 14, 1997.

Ginetz, R. J. Letter to [redacted], Executive Director, B.C. Salmon Farmers Association. RE: Atlantic salmon importation policy. February 17, 1992.

Glass, R. Letter from R. Glass to A. Morton, Raincoast Research at Simoom Sound. December 28, 1994.

Godley, M. Letter from M. Godley, Processing Section to A. Morton, Raincoast Research at Simoom Sound. RE: Applications Pursuant to the Waste Management Act on behalf of Omega Salmon Group Ltd. and Seven Hills Aquafarm Ltd., dated June 24, 1996, located at various sites in the Queen Charlotte Strait area. August 20, 1996.

Godley, M. Letter from M. Godley, Processing Section to A. Morton, Raincoast Research at Simoom Sound. RE: Applications pursuant to the Waste Management Act on behalf of Stolt Sea Farm Inc., dated November—December 1995, located at various sites. March 19, 1996.

Goodrich, S. Letter to Premier Mike Harcourt from the Campbell River Environmental Council. RE: Aquaculture Regulation. December 14, 1992.

Goodrich, S. Letter from S. Goodrich on behalf of Campbell River Environmental Council to J. Crosbie, Minster of Fisheries and Oceans. RE: Importation of Atlantic salmon eggs into British Columbia. May 4, 1993.

Guimont, F. Letter from F. Guimont, Assistant Deputy Minister at environmental Protection Service to A. Morton, Raincoast Research at Simoom Sound. RE: Feed. February 2, 1999.

Hackett, B. Letter to Pat Chamut Director General Pacific Region, Fisheries and Oceans RE: Federal-Provincial policy. RE: Importation of live salmonids. December 18, 1986.

Hackett, B.A. Letter to P. Chamut, Director General of Pacific Region, Department of Fisheries and Oceans and D. Narver, Director, Fisheries Branch BC Ministry of Environment and Parks. RE: Federal-Provincial policy for the importation of live salmonids into British Columbia. October 17, 1986.

Harcourt, M. Letter to S. Goodrich, Campbell River Environmental Council. RE: Infectious haematopoetic necrosis (IHN) in farmed salmon. January 25, 1993.

Hastein, T. Letter from T. Hastein to A. Morton, Raincoast Research at Simoom Sound. May 31, 1995.

Hoskins, G. Letter to Dr. B. Scott IBEC. RE: Atlantic salmon importation from Scotland quarantine facility design and operation. August 27, 1984.

Hoskins, G. [Redacted letter] to B.C. Packers Limited. RE: Atlantic salmon egg imports. January 2, 1991.

Hoskins, G. Letter to P. Chamut, Director General, Pacific Region, Department of Fisheries and Oceans from BC Environment Lands and Parks. RE: Opposition to the import of Atlantic salmon eggs. November 29, 1991.

Hoskins, G. Letter to [redacted]. RE: Remote risk to diseases pathogenic to salmon. October 10, 1993.

Hoskins, G. Letter to [redacted]. RE: Import restrictions and the fish farmer. October 14, 1986.

Johns, D. Letter from D. Johns, Director of Land Use Branch to A. Morton, Raincoast Research at Simoom Sound. April 15, 1996.

Johnson, T.R. Letter to Tom May B.C. Salmon Farmers Association. RE: Federal-Provincial salmonid import policy. April 24, 1987.

Jones, G. Letter to J.C. Davis. RE: Federal/Provincial policy—Importation of salmonids to British Columbia. April 22, 1987.

Joone, G.C. Letter to D. Swan at Sea Farms Technology. RE: Kennedy Lake hatchery proposal. November 29, 1985.

Kangasniemi, B. Letter from B. Kangasniemi, Biologist at Water Quality Branch to A. Morton, Raincoast Research at Simoom Sound. September 30, 1993.

Kieser, D. Letter from D. Kieser, Fish Pathologist to A. Morton, Raincoast Research at Simoom Sound. RE: Your letter of January 30, 1998. April 28, 1998.

Kieser, D. Letter to A. Morton, Raincoast Research. RE: Furunculosis at the Scott Cove Hatchery. February 1, 1993.

Kieser, D. Letter from D. Kieser, A/Head, Fish Pathology Program at Pacific Biological Station to A. Morton, Raincoast Research at Simoom Sound. RE: Scot Cove. February 10, 1995.

Kieser, D. Letter from D. Kieser, Fish Pathologist to A. Morton, Raincoast Research at Simoom Sound. RE: Formalin fixed kidney samples. July 21, 1993.

Kieser, D. Letter from D. Kieser to A. Morton, Raincoast Research at Simoom Sound. June 26, 1995.

Kieser, D. Letter from D. Kieser, Fish Pathologist to A. Morton, Raincoast Research at Simoom Sound. March 30, 1995.

Kieser, D. Letter from D. Kieser, Fish Pathologist to A. Morton, Director Raincoast Research at Simoom Sound. March 8, 1993.

Kieser, D. Letter from D. Kieser, Fish Pathologist to A. Morton, Raincoast Research at Simoom Sound. RE: Absence of infectious salmon anemia in B.C. November 17, 1997.

Kieser, D. Letter from D. Kieser, Fish Pathologist to A. Morton, Raincoast Research at Simoom Sound. RE: Strains *of Aeromonas salmonicida* found in Scott Cove stocks. November 21, 1996.

Knowles, T. Letter from T. Knowles, Governor to G. Campbell. April 30, 2002.

Koch, A. Letter from A. Koch, Economic Development Officer to A. Morton, Raincoast Research at Simoom Sound. RE: Whales. June 10, 1996.

Lawrie, J. Letter to G. Hoskins. RE: Atlantic salmon: Future egg imports from Sea Farm Canada, Inc. February 13, 1989.

Lawrie, J. Letter to R. Ginetz, Department of Fisheries and Oceans from Sea Farm Canada Inc. RE: Request for the importation of eyed Atlantic salmon ova for the genetic enhancement of farmed stock. July 19, 1990.

Lehmann, B. Letter to A. Wood, Department of Fisheries and Oceans, Western Trout Farmer's Association. RE: Against introduction of Atlantic salmon importation. May 19, 1987.

Lindbergh, J. Letter to U.S Trade Representative Office. RE: Wash Growers position. December 2, 1991.

Lindbergh, J. Letter to D. Narver, BC Ministry of Environment and Parks from Washington Fish Growers Association. RE: Importation of Atlantic salmon smolts and eyed eggs into British Columbia from the United States. July 13, 1992.

Lindbergh, J.M. Letter to Don Tillapaugh, B.C. Salmon Growers Association. RE: Importation of Atlantic salmon smolt and eyed eggs into British Columbia from the United States. May 19, 1992.

Lochbaum, E. Letter from E. Lochbaum, A/Marine Mammal Coordinator to A. Morton, Raincoast Research at Simoom Sound. RE: Acoustic deterrent devices. September 29, 1998.

Ludwig, B. Letter from B. Ludwig, Head Biological Support Unit at Fish Culture Section to A. Morton, Raincoast Research at Simoom Sound. RE: Chinook samples from Kingcome Inlet. August 30, 1994.

Ludwig, B. Letter from B. Ludwig, Head Biological Support Unit to A. Morton, Raincoast Research at Simoom Sound. RE: Escapes of farmed fish. July 27, 1993.

Ludwig, B. Letter from B. Ludwig, Head of Biological Support Unit at Fish Culture Section to A. Morton, Raincoast Research at Simoom Sound. RE: Chile. November 12, 1993.

Macbride, L. Letter to S. Collision, Aquaculture and Commercial Fisheries Branch, Ministry of Agriculture, Fisheries and Food from Save Georgia Strait Alliance. RE: Recommendations of Minister's Aquaculture Industry Advisory Council. January 21, 1993.

Marr, B.C. Letter to P. Meyboom, Deputy Minister Department of Fisheries and Oceans. RE: Draft of the Federal-Provincial policy for the importation of live salmonids into British Columbia. February 2, 1987.

Mayser, R. Letter from R. Mayser, manager at Crown Lands and Resources to A. Morton, Raincoast Research at Simoom Sound. September 17, 2010.

McColl, D. Letter to Alexandra Morton. RE: Broughton C.R.I.S. study. April 27, 1990.

McKay, J. Letter to Pat Chamut. RE: Approval for importation of Atlantic eyed eggs from United Hatcheries Ltd. July 27, 1990.

McKay, J. Letter to R. Ginetz DFO Aquaculture Coordinator. RE: Approval for importation of Atlantic eyed eggs from United Hatcheries Ltd. July 27, 1990.

McKay, J. Letter to R.J. Ginetz, Chief, Aquaculture Division, Fisheries Branch, Pacific Region, Department of Fisheries and Oceans, from United Hatcheries. RE: Joint Venture Multi Company Atlantic Quarantine Program. May 31, 1991.

McKay, J.A. Letter to G. Hoskins, Fish Health Official, Department of Fisheries and Oceans from United Hatcheries Ltd. RE: 1992–93 Quarantine Atlantic eyed egg inventory. August 27, 1992.

McKinnel, S. Letter to A. Morton, Raincoast Research. RE: Atlantic salmon escapes. February 16, 1993.

McKinnell, S. Letter to S. McKinnell, Pacific Biological Station. RE: Escaped farm salmon. February 24, 1993.

McLelland, N. Letter from N. McLelland, Director of Access to Information and Privacy to A. Morton. June 1, 2010.

Mifflin, F. Letter from F. Mifflin, Rear-Admiral to A. Morton, Raincoast Research at Simoom Sound. RE: Limits. September 25, 1996.

Morton, A. Letter to T.D. Stirling, School of Computing Science, Simon Fraser University. RE: Sores on wild salmon around fish farms. 1993.

Morton, A. Letter from A. Morton to M.M. Sexton, Environmental Protection Officer. April 22, 1996.

Morton, A. Letter to J. Walker, Assistant Deputy Minister, Ministry of Environment, Lands and Parks. RE: Fish farms and Kingcome Inlet. April 26, 1993.

Morton, A. Letter from A. Morton to B. Tobin, Minister of Fisheries and Oceans. August 10, 1994.

Morton, A. Letter to M. Nock. RE: Opposition to fish farm sites in Queen Charlotte Strait. August 18, 1993.

Morton, A. Letter to Department of Fisheries and Oceans. August 25, 1993.

Morton, A. Letter to J. Lewis, Department of Fisheries. RE: Wary of the salmon farming industry. February 11, 1993.

Morton, A. Letter from A. Morton to D. Kieser, Pacific Biological Station. February 18, 1995.

Morton, A. Letter from A. Morton to D. Anderson, Fisheries Minister. February 20, 1999.

Morton, A. Letter to D. Kieser, Pacific Biological Station. RE: Aeromonas salmonicida resilience. February 24, 1993.

Morton, A. Letter to T. Smith, Pacific Biological Station. RE: Fish farm permit conditions. February 24, 1993.

Morton, A. Letter from A. Morton to H. Andrusak, Ministry of Environment. RE: Atlantics spawning in wild. February 9, 1994.

Morton, A. Letter to J. Boutillier, Head of Aquaculture Division, Department of Fisheries and Oceans. RE: Fish farms impacts. January 12, 1992.

Morton, A. Letter from A. Morton to B. Tobin. RE: Response. January 13, 1995.

Morton, A. Letter from A. Morton to D. Deans, Director Habitat Management at Pacific Region, DFO. July 14, 1993.

Morton, A. Letter from A. Morton to G. Caine, Chief of Field Operations Aquaculture & Commercial Fisheries Branch, Ministry of Agriculture and Fisheries. July 15, 1993.

Morton, A. Letter from A. Morton to D. Petrachenko and Pacific Region. RE: Wild stocks collapsing. July 25, 1997.

Morton, A. Letter from A. Morton to E. Cull and B. Tobin. July 30, 1995.

Morton, A. Letter to D. Narver. RE: Wild/farmed salmonid interactions. July 6, 1993.

Morton, A. Letter from A. Morton to J. Cashore, Minister of Environment. June 23, 1993.
Morton, A. Letter from A. Morton to A. Castledine, M.A.F.F. June 28, 1993.
Morton, A. Letter from A. Morton to J. Sawicki, Minister of Environment. March 12, 2000.
Morton, A. Letter from A. Morton to H. Andrusak, Manager Fish Culture Section. March 29, 1993.
Morton, A. Letter from A. Morton to Minister Sihota. March 5, 1995.
Morton, A. Letter from A. Morton to D. Anderson, Minister of Fisheries. March 7, 1998.
Morton, A. Letter from A. Morton to Minister Tobin. March 8, 1993.
Morton, A. Letter from A. Morton to Minister Tobin, Minister of Fisheries. March 8, 1995.
Morton, A. Letter from A. Morton to P. Chamut, Regional Head of DFO. May 10, 1993.
Morton, A. Letter from A. Morton to R. Ginetz. May 10, 1993.
Morton, A. Letter from A. Morton to the Ministry of Environment, Lands and Parks. May 24, 1993.
Morton, A. Letter from A. Morton to Ted and Nora Stirling, School of Computer Sciences, SFU. May 29, 1993.
Morton, A. Letter from A. Morton to Ministry of Environment, Lands and Parks. May 4, 1993.
Morton, A. Letter from A. Morton, Raincoast Research at Simoom Sound to B. Tobin, Minister of Fisheries. November 11, 1994.
Morton, A. Letter from A. Morton to J. Walker, Assistant Deputy Minister at Ministry of Environment. November 14, 1995.
Morton, A. Letter from A. Morton, Raincoast Research at Simoom Sound to J. Walker. November 4, 1994.
Morton, A. Letter from A. Morton, Raincoast Research at Simoom Sound to M. Sihota. November 8, 1994.
Morton, A. Letter from A. Morton to D. Petrachenko, Director-General of Pacific Region. October 15, 1997.

Morton, A. Letter from A. Morton to D. Petrachenko, Director-General of Pacific Region. October 15, 1997.

Morton, A. Letter from A. Morton, Raincoast Research at Simoom Sound to Dr. F. Utter, University of Washington School of Fisheries. October 2, 1996.

Morton, A. Letter from A. Morton to D. Keiser, Pacific Biological Station. October 26, 1997.

Morton, A. Letter from A. Morton to M. Sihota. September 1, 1995.

Morton, A. Letter from A. Morton to J. Walker, Assistant Deputy Minister at Ministry of Environment. September 18, 1995.

Morton, A. Letter from Alex Morton to Ministry of Environment. May 1993.

Morton, A. Letter from A. Morton to P. Christie, M.E.L.P. BC Lands. [undated].

Mulholland, M. Letter to G. Hunter. RE: Import of Atlantic salmon eggs from Norway and Scotland. October 30, 1985.

Narver, D. Letter to A. Morton. RE: Response to July 6, 1993 letter. August 11, 1993.

Narver, D. Letter to P. Chamut, Director General Pacific Region, Department of Fisheries and Ocean. RE: Opposition to changing Federal-Provincial policy on the import of Atlantic salmon eggs. December 6, 1991.

Narver, D. Letter from D. Narver, Director to P. Chamut, Director General Pacific Region at Department of Fisheries and Oceans. RE: Atlantic salmon egg import policy. March 29, 1993.

Narver, D. Letter from D. Narver, Director to P. Chamut, Director General of Department of Fisheries and Oceans. RE: Non-reproductive chinook. May 25, 1994.

Narver, D. Letter to T.A. Watson, Pacific Aqua Foods Ltd, RE: Crystal Waters exemption to allow import of 750,000 eggs. May 29, 1987.

Narver, D. Letter to Dale Blackburn Sea Farm Canada Inc. RE: Exception to regulation ask and risk to wild salmon. November 7, 1986.

Narver, D. Letter to Assistant Deputy Minister, Ministry of Agriculture and Fisheries, B.A. Hackett. November 7, 1986.

Narver, D. Letter to J.M. Lindbergh, Washington Fish Growers Association. RE: Importation of Atlantic salmon. September 16, 1992.

Needham, T. Letter to G. Hoskins RE: Import of Atlantic salmon eggs into British Columbia from BC Packers Limited. December 3, 1990.

Noakes, D. Letter to Professor Cho-Teng Liu. March 1999.

Nock, M. Letter to A. Morton. RE: Response to June 23, 1993. August 5, 1993.

Nock, M. Letter to M. Nock, Ministry of Environment Lands and Parks. RE: Salmon farm application. August 28, 1993.

O'Riordan, J. Letter from J. O'Riordan, Assistant Deputy Minister at Environment and Lands Regions Division to A. Morton, Raincoast Research at Simoom Sound. March 1, 2000.

Petrachenko, D. Letter from D.M. Petrachenko, Regional Director General of Pacific Region to A. Morton, Raincoast Research at Simoom Sound. February 24, 1999.

Petrachenko, D. Letter from D.M. Petrachenko, Director-General of Pacific Region to A. Morton, Raincoast Research at Simoom Sound. RE: No evidence. January 18, 1999.

Petrachenko, D. Letter from D.M. Petrachenko, Director-General of Pacific Region to A. Morton, Raincoast Research at Simoom Sound. RE: Fish health issues. January 18, 1999.

Petrachenko, D. Letter from D.M. Petrachenko, Director-General of Pacific Region to A. Morton, Raincoast Research at Simoom Sound. RE: Evidence farm fishing is good. July 6, 1997.

Petrachenko, D. Letter from D. Petrachenko, Director-General of Pacific Region to A. Morton, Raincoast Research at Simoom Sound. October 1, 1997.

Petrachenko, D. Letter from D. Petrachenko, Director-General of Pacific Region to A. Morton, Raincoast Research at Simoom Sound. October 1, 1997.

Price, I. Letter to [redacted], Washington Fish Growers Association. RE: B.C. Salmon Farmers Association importation position. September 4, 1992.

Ptolemy, R.A. Letter to A. Morton. RE: Request for publications and manuscript. August 24, 1993.

Reeder, S. Letter from BC Wildlife to Erik Neilson Acting Minister of Fisheries and Oceans. November 15, 1985.

Rock, A. Letter from A. Rock, Ministry of Health to A. Morton, Raincoast Research at Simoom Sound. August 25, 2000.

Russell, M. Letter from M. Russell to B. Tobin, Minister of Fisheries and Oceans. March 18, 1994.

Saunders, R. Letter from R. Saunders to A. Morton, Raincoast Research at Simoom Sound. January 11, 1994.

Saunders, R. Letter from R.L. Saunders to R.J. Gibson, Science Branch at Department of Fisheries and Oceans. RE: Alex's paper. November 24, 1993.

Schmitten, R. Letter to P. Chamut from United States Department of Commerce. RE: Atlantic salmon import protocol. November 21, 1991.

Scottish Office. Letter from the Scottish Office, Agriculture and Fisheries Department to A. Morton, Raincoast Research at Simoom Sound. March 29, 1995.

Sexton, M. Letter from M. Sexton, Environmental Protection Officer of Municipal Section to A. Morton, Raincoast Research at Simoom Sound. RE: Application Pursuant to the Waste Management Act on behalf of Stolt Sea Farm Inc., dated November—December 1995, located at various sites. May 3, 1996.

Siddon, T. Letter to Co-op Fisherman's Guild [redacted]. RE: Importation of disease via Atlantic salmon eggs. April 25, 1986.

Sihota, M. Letter from M. Sihota, Minister to A. Morton, Raincoast Research at Simoom Sound. April 6, 1995.

Sihota, M. Letter from M. Sihota, Minister to A. Morton, Raincoast Research at Simoom Sound. February 13, 1996.

Sihota, M. Letter from M. Sihota, Minister to A. Morton, Raincoast Research at Simoom Sound. January 19, 1996.

Sihota, M. Letter from M. Sihota, Minister to A. Morton, Raincoast Research at Simoom Sound. RE: Workshop attendance. January 23, 1995.

Sihota, M. Letter from M. Sihota, MLA (Esquimalt-Metchosin) to A. Morton, Raincoast Research at Simoom Sound. RE: Kind letter. June 20, 1995.

Smith, D. Letter from D. Smith, Barrister and Solicitor to W. Smart, Barristers and Solicitors. RE: Morton, Alexandra v. Heritage Salmon Ltd. et al. Provincial Court File No. 13381-1. Port Hardy Registry. December 28, 2005.

Smith, T. Letter to A. Morton, Director, Raincoat Research from Pacific Biological Station, Department of Fisheries and Oceans. RE: Predator licences and firearm safety. February 1, 1993.

Souza, L. Letter to F. Carpenter, Program Officer—Aquaculture Division, Department of Fisheries and Oceans. RE: Farmed Atlantic salmon in British Columbia. August 1, 1990.

Souza, L. Letter to F. Carpenter, Program Officer, Aquaculture Division, Fisheries Branch—Pacific Region. RE: Sea Farm Canada In—Land Ref: 1406187. June 6, 1990.

Stechy, D. Letter from D. Stechy, Director of Aquaculture to B. Proctor, Simoom Sound. September 30, 1993.

Stewart, J. Letter from J. Stewart, Habitat Ecology Division of Biological Science Branch to A. Morton, Raincoast Research at Simoom Sound. April 10, 1995.

Stewart, J. Letter from J. Stewart, Habitat Ecology Division to A. Morton. February 21, 1995.

Stewart, J. Letter from J. Stewart, Habitat Ecology Division to A. Morton, Raincoast Research at Simoom Sound. February 22, 1995.

Stewart, J. Letter from J. Stewart to A. Morton, Raincoast Research at Simoom Sound. March 15, 1995.

Stewart, J. Letter from J. Stewart, Habitat Ecology Division to A. Morton, Raincoast Research at Simoom Sound. RE: Furunculosis. March 24, 1995.

Stewart, J. Letter from J. Stewart, Habitat Ecology Division to A. Morton, Raincoast Research at Simoom Sound. March 3, 1995.

Streifel, D. Letter from D. Streifel, Minister to A. Morton, Raincoast Research at Simoom Sound. February 23, 2000.

Thibault, R. Letter from R. Thibault to A. Morton, Raincoast Research at Simoom Sound. December 10, 2002.

Thomson, A. Letter from A.J. Thomson, Contracted Biologist to A. Morton, Raincoast Research at Simoom Sound. RE: Samples. February 9, 1994.

Thomson, A. Letter from A. Thomson, Contracted Biologist to A. Morton, Raincoast Research at Simoom Sound. RE: Stomach Contents. November 2, 1994.

Thomson, A. Letter from A.J. Thomson, Contract Biologist to A. Morton, Raincoast Research at Simoom Sound. RE: Sockeye salmon sampling. October 16, 1997.

Tillapaugh, D. Letter to J. Walker Clayoquot Sound Sustainable Development secretariat from the B.C. Salmon Farmers Association. RE: Support of lake pen salmon smolts in Kennedy Lake. April 16, 1992.

Tillapaugh, D. Letter to Ron Ginetz, Chief Aquaculture Division, Pacific Region, Department of Fisheries and Oceans from the BC Salmon Farmers Association. RE: Alterations to the import policy for Atlantic eggs and smolts (in particular from Washington State). January 14, 1992.

Tillapaugh, D. Letter to Jon Lindbergh, Washington State Fish Growers Association. RE: Importation of Atlantic salmon smolts and eyed eggs into British Columbia from the United States. January 15, 1992.

Tillapaugh, D. Letter to R. Ginetz, Chief, Aquaculture Division, Pacific Region, Department of Fisheries and Oceans from the B.C. Salmon Farmers Association. RE: Association position on Atlantic salmon import policy. July 30, 1992.

Tobin, B. Letter from B. Tobin, Minister of Fisheries to A. Morton. RE: Fish farming doesn't pose a threat. February 2, 1995.

Tobin, B. Letter from B. Tobin to A. Morton, Raincoast Research at Simoom Sound. RE: No firm evidence. July 26, 1994.

Tobin, B. Letter from B. Tobin, Minister of Fisheries and Oceans to I. Novaczek, Canadian Oceans Caucus. May 10, 1994.

Tobin, B. Letter from B. Tobin to A. Morton, Raincoast Research at Simoom Sound. RE: No harm. November 9, 1994.

Tobin, B. Letter from B. Tobin to A. Morton, Raincoast Research at Simoom Sound. RE: Acoustic deterrent. November 9, 1995.

Tobin, B. Letter from B. Tobin to A. Morton, Raincoast Research at Simoom Sound. RE: Does not pose a threat. September 9, 1994.

Tousignant, L. Letter from L. Tousignant, Director-General of Pacific Region to A. Morton, Raincoast Research at Simoom Sound. RE: Acoustic. February 26, 1996.

Tousignant, L. Letter from L. Tousignant, Director-General of Pacific Region to A. Morton, Raincoast Research at Simoom Sound. RE: Your concerns regarding importation of Atlantic salmon eggs. January 10, 1996.

Tousignant, L. Letter from L. Tousignant, Director-General of Pacific Region to A. Morton, Raincoast Research at Simoom Sound. RE: Strict egg policy. September 26, 1996.

Van Dongen, J. Letter from J. Van Dongen, Minister to A. Morton, Raincoast Research at Simoom Sound. December 10, 2002.

Walker, J. Letter to A. Morton. RE: Response to February 17, 1993 letter. April 19, 1993.

Walker, J. Letter from J.H.C. Walker, Assistant Deputy Minister to A. Morton, Raincoast Research at Simoom Sound. May 17, 1993.

Walker, J. Letter from J.H.C. Walker, Assistant Deputy Minister at Fisheries, Wildlife and Habitat Protection Department to A. Morton, Raincoast Research at Simoom Sound. November 23, 1995.

Walker, J. Letter from J. Walker, Assistant Deputy Minister to A. Morton, Raincoast Research at Simoom Sound. September 5, 1995.

Walker, J. Letter from J. Walker, Assistant Deputy Minister to A. Morton, Raincoast Research at Simoom Sound. January 9, 1996.

Watson, T. Letter to D. Narver, Ministry of Environment and Parks. RE: Limitation of Atlantic salmon eggs for the Crystal Waters Hatchery. May 26, 1987.

Wynn, S. Letter from Dr. S. Wynn, Deputy Minister to A. Morton, Raincoast Research at Simoom Sound. RE: Fish farm review. March 17, 1998.

MEMORANDA

Anon. U.S. Response to Canadian Protocol on importation of Atlantic salmon eggs and smolts, U.S.—Canada Bilateral Meeting of the Working Group on Fish and Fish Products Inspection. December 20, 1991.

Anon. Discussion document [internal memorandum]: National Policy on Introduction and Transfers of Aquatic Organisms. April 5, 1993.

Beamish, D. Memorandum to P. Chamut, Director General Pacific Region, Department of Fisheries and Oceans. RE: Request for approval for the aquaculture industry to import Atlantic salmon eggs into BC under Section 4 of the *Fisheries Act*. December 19, 1990.

Beamish, D. Memorandum to P. Chamut, Director General Pacific Region, Department of Fisheries and Oceans. RE: Atlantic salmon imports into British Columbia. January 3, 1991.

Boutillier, J. Memorandum to J.C. Davis, R.J. Beamish, and R. Ginetz. RE: Protocol for B.C. on Atlantic salmon imports. February 7, 1992.

Carey, T. Interoffice memorandum to T. Tebb and R. Ginetz. RE: Follow-up meeting with USA—B.C. policy on Atlantic salmon. December 30, 1991.

Carey, T. Memorandum from T. Carey, Aquaculture and Oceans Science Branch to local fish health officers/agents. RE: Province-by-province policies on allowing like-to-like transfers under the Amended Fish Health Protection Regulations FHPR. March 19, 1997.

Chamut, P. Memorandum from P. Chamut, Director General of Pacific Region to Dr. W.G. Doubleday, A/ Assistant Deputy Minister of Science Sector. RE: Atlantic salmon importation policy—Trade issue with U.S.A. March 23, 1992.

Chamut, P. Memorandum from P. Chamut, Director General of Pacific Region to L.S. Parsons, Assistant Deputy Minister of Science. RE: Policy on importation of salmonids. March 23, 1992.

Chamut, P. Memorandum to D. Narver. RE: Federal-provincial policy on Atlantic salmon importation. C. 1988.

Chamut, P. Memorandum to J. Davis, Regional Director. RE: Restrictions on import of Atlantic salmon. December 19, 1990.

Chamut, P. Memorandum to D. Good and B. Morrissey. RE: Farmed Atlantic salmon in B.C. April 23, 1991.

Chamut, P. Memorandum to J.C. Davis. RE: Atlantic salmon importation policy. January 15, 1992.

Chamut, P. Memorandum to B. Rawson. RE: Atlantic salmon import policy. February 28, 1992.

Chamut, P. Policy for the importation of Atlantic salmon into British Columbia. [Internal memorandum]. March 2, 1992.

Davis, J. Memorandum to P. Chamut, Regional Director General Pacific Region. RE: Importation of salmonid eggs into British Columbia your note of June 27/87—copy attached. July 23, 1987.

Davis, J.C. Memorandum to G.E. Jones, Regional Director, Fisheries Branch, Pacific Region. RE: Federal/provincial policy—importation of salmonids to British Columbia. June 17, 1987.

Davis, J.C. Memorandum to G.E. Jones, Regional Director, Fisheries Branch, Science. RE: Federal/provincial policy—importation of salmonids to British Columbia. May 26, 1987.

Davis, J. Memorandum to P. Chamut, Director General Pacific Region, Fisheries and Oceans. RE: Importation of non-salmonid fish into British Columbia. August 16, 1990.

Doubleday, W.G. Memorandum to P. Chamut. RE: Trade Issue with USA—B.C. Atlantic salmon policy. July 30, 1992.

Emberely, B.J. Memorandum to Director, Aquaculture and Resource Development Branch, Department of Fisheries and Oceans. June 20, 1991.

Fralick, J. Memorandum to T. Halsey, Manager Marine Resources Section. RE: PBS Broodstock Program. April 12, 1985.

Ginetz, R. Canadian response to USA concerns with the policy on importation of Atlantic salmon to British Columbia. March 16, 1992.

Ginetz, R. Memorandum from R.M.J. Ginetz to H. Andrusak, Ministry of Environment. C. August 1991.

Ginetz, R. Memorandum to A.F. Lill. RE: Aquaculture issues—Direction/strategy. February 19, 1991.

Ginetz, R.J. Memorandum to A.F. Lill. RE: Federal-provincial policy on the importation of Atlantic salmon into B.C. June 11, 1991.

Ginetz, R.J. Memorandum to I. Price. RE: Atlantic salmon importation policy. July 21, 1992.

Ginetz, R. Answers to External Affairs Questions—Memo of February 27, 1992. March 3, 1992.

Ginetz, R. Memorandum from R. Ginetz, Regional Aquaculture Coordinator of Pacific Region to G. Goulet. February 3, 1997.

Ginetz, R. Memorandum from R. Ginetz, Regional Aquaculture Coordinator of Pacific Region to P. Redford, Finn Bay Seaproducts Ltd. February 17, 1997.

Ginetz, R. Memorandum from R. Ginetz, Regional Aquaculture Coordinator of Pacific Region to E. Downey, Paradise Bay Seafarms. February 24, 1997.

Hoskins, G. Memo to W Shinners, Director General Pacific Region RE: Atlantic salmon imports by Ibec. April 29, 1985.

Hoskins, G. Memorandum to M.J. Comfort. RE: Atlantic salmon eggs from Scotland. December 10, 1985.

Hoskins, G. Memorandum to T. Tebb. RE: Atlantic salmon imports and the discovery of viral hemorraghic septicemia in Washington state. July 4, 1991.

Hoskins, G. Memorandum to P. Chamut. RE: Importation of Atlantic salmon eggs into B.C. Under Section 4 of the Fisheries Act. November 25, 1991.

Hunter, G. Memorandum to F.E.A. Wood and R.J. Beamish. RE: Termination of Atlantic egg imports into B.C. July 9, 1985.

Kieser, D. Memorandum to J.C. Davis from Federal/Provincial Transplant Committee. RE: Proposed protocol and guidelines for importations of finfish and shellfish into British Columbia. January 7, 1991.

Kieser, D. Memorandum from D. Kieser, Fish Pathologist at Pacific Biological Station to D. Knierem, Manager Scott Cove Hatchery and M. Berry, Technical Advisor, and G. Bates, Community Advisor at Scott Cove Hatchery. RE: Disease testing of 1993 coho broodstock, Scott Cove Hatchery. March 14, 1994.

Kieser, D. Memorandum from D. Kieser, Fish Pathology Program to B. Proctor, Scott Cove Hatchery. RE: Scott Cove broodstock—furunculosis control in 1995. August 2, 1995.

Kieser, D. Memorandum from D. Kieser to D. Noaks. RE: Comments on Net Loss report by David Ellis and Associates. November 4, 1996.

Kieser, D. Memorandum from D. Kieser to D. Noakes. RE: Meeting. April 1, 1997.

Kieser, D. Memorandum from D. Noakes to D. Kieser. RE: Up-dating of salmonid importation policies. April, 16, 1997.

Ministry of Environment Lands and Parks. Furunculosis bacteria resistant to the antibiotic erythromycin. [Internal memorandum]. Province of British Columbia. December 9, 1993.

Muir, B. Memorandum to P. Chamut, Director General Pacific Region, Fisheries and Oceans. RE: Importation of Atlantic salmon eggs under Section 4, Fisheries Act. December 20, 1990.

Narver, D. Memorandum to E.D. Anthony, Assistant Deputy Minister. RE: 1985 RE: Imports of Atlantic salmon eggs. February 26, 1985.

Narver, D. MOU between P. Chamut Director General Pacific Region, Department of Fisheries and Oceans and D. Narver, Director, Recreational Fisheries Branch, Ministry of Environment and Parks. June 8, 1987.

Narver, D., and P. Chamut. Policy memorandum of understanding. May 3, 1988.

Narver, D., and P. Chamut. Memorandum RE: Federal—provincial policy for the importation of Pacific salmon into British Columbia regarding Stikine sockeye. June 8, 1989.

Parsons, L. Memorandum to P. Chamut Director General Pacific Region. RE: Policy on importation of salmonids. June 9, 1987.

Price, I. Interoffice Memorandum. RE: Teleconference with USA RE: Salmon eggs. February 18, 1992.

Price, I.M., and J.G. Nickum. Memorandum to B.J. Emberley, Department of Fisheries and Oceans and T.J. Billy, U.S Department of Commerce. RE: Working Group on Fish and Fishery Products Inspection. August 27, 1991.

Price, I. Memorandum to J. Emberley, Director General Inspection Services. RE: B.C. Policy on Atlantic and Pacific salmon importations. August 30, 1991.

Price, I. Memorandum to R. Ginetz. RE: Atlantic salmon eggs and US Department of Commerce. November 26, 1991.

Price, I. Memorandum to K. Roeske. RE: Pacific Region policy on importation of Atlantic salmon. March 19, 1992.

Roeka, K., and A. Sarna. Memorandum to WSHDC. RE: Fish inspection issues, salmon egg import policy and Canadian side explained rationale. July 16, 1991.

Russell, R. Memo to Paul Sprout, Private Salmon Hatcheries and Netpen Facilities—Some Serious Concerns. Government of Canada. [Internal memorandum]. August 11, 1988.

Siddon, T., and J. Savage. Canada/British Columbia memorandum of understanding on aquaculture development. September 6, 1988.

Miscellaneous Publications

Breathnach, M., Whelan, K., and Piggins, D. (1992, February 18). STAG pinpoints role of sea lice in sea trout collapse. Stag Sea Trout Action Group Press Release.

Cummins, J. (2002, March 5). Senator Murkowski warns of the dangers of salmon feedlot expansion. Official Opposition Critic Fisheries and Oceans Press Release.

Cummins, J. (2002, January 31). DFO unready and unwilling to handle salmon farm expansion sports and commercial industry put at risk. Official Opposition Critic Fisheries and Oceans News Release.

Fleming, I. (1997, October 3). Fax from I. Fleming to D. Ellis.

Hinkson, C. (February 9, 2009). "The Morton Decision." *Morton v. British Columbia (Agriculture and Lands),* **2009 BCSC 136.**

Jones, J.E. (December 23, 1985). Scientific experimental permit for Seafarms Technology Ltd.

Meyboom, P., and Maar, B. (1987). Federal-Provincial policy for the importation of live salmonids into British Columbia. Draft policy document.

Mills, E. (February 7, 2006). "Fisheries Research Board." The *Canadian Encyclopedia.*

Morton, A. (August 23, 2010). Fish farms operating without valid Crown land tenures: Biologist applies for expired salmon feedlot licenses. Press release.

Price, I., and Carey, T. (July 26, 1991). Summary draft RE: Atlantic salmon import policies.

Province of British Columbia. (December 9, 1993). Minutes of the joint DFO, MAFF and MELP meeting (December 9, 1993) to discuss information, implications and action arising out of the recent isolation of Furunculosis bacteria resistant to the antibiotic erythromycin. Ministry of Environment, Lands and Parks.

R., B. (November 22, 1991). Draft briefing notes on Imports for Deputy RE Introduction of Atlantic salmon eggs into British Columbia—A revision of policy.

Save Our Sea Trout. (August 10, 1992). Sea trout—Why the salmon growers are talking complete rubbish. Press release.

Sundquist, L. (January 14, 2000). Fax from L. Sundquist to L. MacBride, Georgia Strait Alliance RE: Tenure List.

Washington Fish Growers. (n.d.). Draft response of British Columbia/Washington Partnership Committee on Fish Health Regulations concerning the need for revision of Canadian federal—provincial policy for the importation of Atlantic Salmon into British Columbia.

Wing, K. (June 5, 1997). Email from K. Wing to AFS RE: Rule: Daily Summary—6/5/97—Part 3 of 3.